6/94

PALAEOETHNOBOTANY
Plants and Ancient Man in Kashmir

PALAEOETHNOBOTANY
Plants and Ancient Man in Kashmir

FAROOQ A. LONE
MAQSOODA KHAN
G.M. BUTH

A.A. BALKEMA / ROTTERDAM
1993

© 1993 COPYRIGHT RESERVED

ISBN 90 6191 944 4

To
Our parents
and
A'tif Farooq (Cosy)

To
Our parents
and
A. (if Farooq it say)

PREFACE

In the process of excavation of archaeological sites one comes across plant matter of one kind or another. Most archaeological plant remains are found in a charred state and are often incomplete for standard systemic study. In the past, these botanical remains received very scant attention. The method of study was then confined to the morphology of botanical materials, which often led to wrong identification and erroneous conclusions. Recent advancement in plant anatomy, microtechnique, and scanning electron microscopy during the last five decades or so has now made it possible to develop a clear and distinct anatomical structure even from minute fragments of plant remains. Thanks to an active band of research workers all over the world, a clear picture of the origin, domestication, and progressive development of agriculture in different parts of the world is fast emerging. We are now in a position to reconstruct the history of botanical science and man-plant relationships of a period for which there is no written record. Of late, archaeobotanical studies have being oriented toward reconstructing past vegetation and palaeoclimates in different parts of the globe.

Palaeoethnobotanists have realised that archaeobotanical studies from the Indian subcontinent, which is very rich in archaeological sites, have not been systematically conducted. We at the botany department of Kashmir University tried to initiate such studies and our intense hardwork has resulted in this book.

The text deals with the palaeoethnobotany of Kashmir and covers a time period from 4500 B.P. to 1000 B.P. The chapter on keys and criteria for identification is based on extensive morpho-anatomical studies of extant plant materials and is primarily meant for future research workers and students interested in plant anatomy, morphology, and systematics. It includes keys to the identification of various species of *Triticum, Hordeum, Avena,* and *Prunus,* and hard and soft woods, and it lists the criteria for identification of weed seeds of about 200 species.

In order to increase awareness about the changes caused by carbonization in proportions and morphology, artificial carbonization of grains and seeds of some cereals and pulses was carried out, the results of these are given in Chapter 4.

Chapter 5 deals with the actual plant evidence from the Kashmir sites and traces and source of various plant species. Succeeding chapters deal with the diffusion of plants into Kashmir, probable uses of plants to the ancient man, changes in vegetation and climate of Kashmir, the state of agricultural and forest economy, and the history of agriculture in Kashmir. In the chapter 'Statistics in Palaeoethnobotany', the possibility of using different statistical parameters in evolving various palaeoethnobotanical and palaeoecological implications is explored and demonstrated with the case study at Semthan. It is hoped that this book will be of help to people interested in ethnobotany, agriculture, forestry, ecology plant morphology and anatomy, palaeoclimate, archaeology, history, and similar subjects.

It is not possible to acknowledge adequately the many sources of aid that have contributed to the development of this book. Information has been drawn from many places, most important of which is the literature created by botanists. The reference citations given at the end are intended as both a source of more detailed information and an acknowledgement to the original investigator. Let me frankly acknowledge that this work would not have been accomplished without the help of Dr. A. R. Naqshi, Taxonomist at the Centre of Plant Taxonomy, University of Kashmir. The help and guidance of Profs. Vishnu-Mittre, D. P. Agrawal, P. Kachroo, M.R. Vijayraghavan, S/Shri R. S. Bisht, G. S. Gaur and R. N. Kaw is gratefully acknowledged. Friends and colleagues, especially Zafar, Shokat Ali, Ashraf, and Naseem, have also been of immense help. The Birbal Sahni Institute of Palaeobotany, in Lucknow, Physical Research Laboratory in Ahmedabad, and Forest Research Institute in Dehra Dun helped in many ways.

Last but not least, we are grateful to University Grants Commission, New Delhi, for financial assistance between 1983 and 1986.

Department of Botany,　　　　　　　　　　　　FAROOQ. A. LONE
University of Kashmir,　　　　　　　　　　　　MAQSOODA KHAN
Srinagar 190006　　　　　　　　　　　　　　　　G. M. BUTH

CONTENTS

Preface		vii
Chapter 1	**Introduction**	1
Chapter 2	**The Sites and Their Environment**	6
	2.1 Burzahom	6
	2.2 Semthan	10
Chapter 3	**Keys and Criteria for Identification**	14
	3.1 Cereals	14
	3.2 Millets	38
	3.3 Pulses	38
	3.4 Horticultural Fruits	38
	3.5 Weed Seeds	45
Chapter 4	**Artificial Carbonization**	98
Chapter 5	**Archaeological Evidence**	106
	5.1 Cereals	106
	5.2 Millets	133
	5.3 Pulses	136
	5.4 Horticultural Fruits	142
	5.5 Weed Seeds	149
	5.6 Woods	153
Chapter 6	**Diffusion of Plants into Kashmir**	195
Chapter 7	**Probable Uses of the Plants Recovered**	199
Chapter 8	**Vegetation, Climate and The Biotic Factor**	201
Chapter 9	**Origin and History of Agriculture**	204
Chapter 10	**State of Economy**	206
	10.1 Agriculture	206
	10.2 Forestry	214
Chapter 11	**Statistics in Palaeoethnobotany**	218
	11.1 Behaviour of Individual Plant Groups	220
	11.2 Chi-Square Analysis	220

	11.3 Intensity of Occupation	231
	11.4 Species Diversity (H), Species Richness (R) and Species Evenness (E)	231
	11.5 Standard Scores (Z)	234
	11.6 Coefficient of Similarity (S)	237
	11.7 Coefficient of Similarity (T)	237
	11.8 Concluding Remarks	239
	Bibliography	242
Appendix I	**Ecological Consideration of Past and Present Vegetation of Kashmir Valley**	258
	Introduction	258
	Pliocene Vegetation of Kashmir	259
	Pleistocene Vegetation of Kashmir	260
	Present Vegetation of Kashmir	263
	Statistical Observations	265
	Concluding Remarks	267
Appendix II	**Kashmir—Ethnobotanic Present**	270
	A. Food Plants	270
	B. Pulses	270
	C. Vegetables	270
	D. Spices and Condiments	271
	E. Narcotics and Beverages	271
	F. Oilseeds	271
	G. Fruits	271
	H. Fibre and Matting	272
	I. Religious Ceremonies and Rituals	272
	J. Medicinal Plants	272
Appendix III	**Soft and Hard Woods of Kashmir**	273
	Soft Woods	273
	Hard Woods	273
	Index	275

CHAPTER 1

INTRODUCTION

Modern man, as revealed by palaeontological evidence, appeared on the earth about 40,000 to 30,000 years B.P. and within a period of 10,000 to 20,000 years spread throughout the world. It was also about this time that humans added fire-making to their cultural heritage. In the course of evolution man marked his progress by evolving a succession of tools in response to his need for mastering the environment, which primarily meant the quest for food. The grinding, polishing, and fabrication of tools gave him an edge over all other animals in the evolutionary struggle. As his cranial capacity increased, he made skilful use of his hands by virtue of which he was able to control his environment. As a result of human hunting a 'Pleistocene overkill' occurred between 15,000 and 12,000 B.P. when many of the large mammals became extinct. Such interactions with hunted animals, along with fire, constitute some of the first major impacts of humans on the ecosytem (Langenheim and Thimann 1982).

By about 10,000 B.P. mankind achieved an important milestone through the development of agriculture and undoubtedly assumed an ecological role without parallel in the history of this planet. From being a hunter and food gatherer, man started becoming a food producer. Through the cultivation of plants man developed a more dependable source of food which led to an ever-increasing growth of population. The sedentary existence associated with cultivation practices also led to the development of village communities. This phase of activity has been termed the 'Neolithic Revolution' (Childe 1957, Barghoorn 1971, Struever 1971, Hutchinson et al. 1977, Moore 1982, 1985, Rindos 1984).

The story of cultivated plants begins with the time and place of domestication, which not only reveals the environment in which this important measure was taken but also throws light on economic and social environments which led to the present stage of our civilization. Our knowledge about the beginnings and progressive development of agriculture has greatly been advanced by the archaeological evidence. The study of ancient grains and other plant materials associated with them help in tracing the history of crops, associated useful plants, and weeds. It also provides historical background to the origin and spread of agriculture.

Not much evidence is available of domesticated cereals earlier than 10,000 years, although archaeologists have reported on the tools used by ancient man for hunting, fishing, and food gathering from excavations of the sites as old as scores of thousands of years (Sankalia 1974). Archaeologically, the last 10,000 years have been divided into three periods: prehistoric, protohistoric, and historic (Chowdhury 1974). The prehistoric period and the earlier part of the protohistoric period have no written record. In the absence of any written record archaeologists have been collecting information on the prehistoric and protohistoric periods by studying remains from excavations of various archaeological sites. The man-plant relationship during the prehistoric and protohistoric periods has been reconstructed mostly on the evidence of plant remains recovered from archaeological excavations.

Progress in the study of archaeological plant remains is presented at the symposia of the International Work Group for Palaeoethnobotany (IWGP). The eighth symposium of the IWGP was held at Nitra, Czechoslovakia, June 19–24, 1989. Studies in palaeoethnobotany include those of Buschan 1895, Neuweiler 1905, Costantin and Bois 1910, Safford 1917, Harms 1922, Gilmore 1931, Yacovleff and Herrera 1934–35, Anderson 1942, Whitaker 1949, 1981, Mangelsdorf and Smith 1949, Helbaek 1952a, 1952b, 1952c, 1953a, 1953b, 1953c, 1954a, 1954b, 1955, 1956, 1958a, 1958b, 1959a, 1959b, 1960a, 1960b, 1960c, 1960d, 1960e, 1961, 1963, 1964a, 1964b, 1965a, 1965b, 1966a, 1966b, 1966c, 1966d, 1966e, 1969, 1970, 1972, Mangelsdorf et al. 1956, Van Zeist and Botteima 1966, Van Zeist and Casparie 1968, 1984, Hopf 1969, Renfrew 1969, 1973, Yarnell 1969, Hillman 1972, 1975, 1978, 1981, 1984a, 1984b, Moore 1975, Hansen and Renfrew 1978, Erichsen-Brown 1979, Hilu et al. 1979, Wendorf et al. 1979, Wendorf and Schild 1984, Korber-Grohne and Peining 1980, Korber-Grohne 1981, Jarriage and Meadow 1980, Botteima 1984, Miller 1984, Stemler and Falk 1980, 1984, Barigozzi 1986, Harlan 1986, Hillman and Davies 1986, Hillman et al. 1986a, 1986b, Hopf 1986, Zohary 1986, Zohary and Hopf 1986, and a series of papers presented at the session on plant domestication at World Archaeological Congress held at Southampton, England, September 1–7, 1986. In India, palaeoethnobotanical studies include those of Chowdhury and Ghosh 1946, 1951, 1952, 1954–55, 1955, 1957, Ghosh 1950, 1961, Ghosh and Lal 1958, 1961, 1962–63, Vishnu-Mittre 1961, 1962, 1968a, 1968b, 1968c, 1969, 1972, 1974, Chowdhury 1963, 1965a, 1965b, 1970, 1974a, 1974b, Rao and Sahi 1967, Vishnu-Mittre and Gupta 1968–69, Buth and Chowdhury 1971, Chowdhury et al. 1971, 1977, Vishnu-Mittre et al. 1971, 1972a, 1972b, Kajale 1974a, 1974b, 1977a, 1977b, 1977c, 1979, Vishnu-Mittre and Savithri 1974, 1975, 1978, 1982, Buth and Bisht 1981, Buth and Kaw 1985, Buth et al. 1982, 1986a, Saraswat 1982, 1985, 1986, Lone et al. 1986a, 1986b, 1987, 1988.

Over the past few years palaeoethnobotanists have made an increasing effort to apply their data to the question of cultural processes like the evolution of cultivation strategies, long-term stability of subsistence strategies, and agricultural intensification (Asch and Asch 1975, Asch et al. 1979, Dennel 1972, Ford 1979, Pearsall 1979, 1980, 1983, 1986, Minnis 1981).

The very location of Kashmir has helped it to secure elements from cultures of various countries falling in central Asia ranging from Neolithic, which is the earliest known culture in Kashmir, to the medieval times. These cultures melted and reacted and the valley became a culture compound of its own type at various stages of human development. In spite of the fact that the valley of Kashmir is rich in archaeological sites (Pant et al. 1982, fig. 1.1), little work has been done on the plant remains from various sites.

In 1983, with the financial support of University Grants Commission, New Delhi, the research project 'Palaeoethnobotanical investigations of plant remains recovered from archaeological sites of Kashmir' commenced. The research programme comprised the examination of charred plant remains from archaeological sites with the main aim of obtaining information on the relations between plants and ancient man in Kashmir. Two sites, namely, Burzahom in Srinagar district and Semthan in Anantnag district, excavated by the Archaeological Survey of India, were selected for detailed study. The time span embraced by the sites is such that information on the plant husbandry from the early Neolithic to the late Iron Age might be expected. Thus, early as well as more advanced stages of agriculture are represented in the study. Since the sites are situated in the same ecological zone, a continuous record of the plant husbandry covering a period of about 3,500 years could be obtained.

The introductory investigations provided a fair impression of the archaeological plant remains, such as diversity of plant species and concentration and preservation of seeds, fruits, and woods. Moreover, experience was gained and collection of modern seed, fruit, and wood reference material could, at least to a certain extent, be directed to particular taxa represented or probably represented in the charred plant record.

The main objects of the study were to obtain information on the following questions:

Was agriculture practised by the ancient inhabitants of Kashmir and, if so, which crops were grown?
Which wild seeds and fruits were collected for human consumption?
What was the vegetation in the valley in the past and how has it changed?
How best did the inhabitants use the natural resources that were available to them in the environment they settled in?

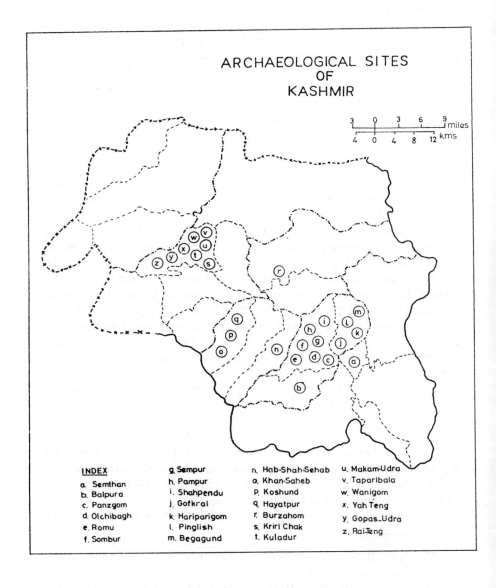

Figure 1.1: Archaeological sites of Kashmir.

What is the origin and source of various plant species recovered and what light do the plant remains throw on the diffusion of plants, probable uses of plants to ancient man, and progressive development of Kashmir agriculture?

How can statistical parameters help in arriving at conclusions regarding palaeoethnobotany and palaeoecology?

The plan adopted here has been to present the criteria for identification based on studies on extant materials, description of archaeological materials, and the archaeobotanical significance of the finds in the light of evidence from other regions of the world.

CHAPTER 2

THE SITES AND THEIR ENVIRONMENT

2.1. BURZAHOM

Burzahom is situated in Srinagar district, about 16 km northeast of Srinagar in Kashmir valley, between Srinagar and Ganderbal. It is located at 34°10′ N latitude and 73°54′ E longitude. The site (fig. 2.1) stands on the ancient lake bed called Karewa (Uddar in Kashmiri dialect) at the foot of the Mahadeva hills and commands a panoramic view of lush, green fields in the immediate vicinity and the shimmering waters of Dal lake, which is hardly 2 km away.

The surrounding lowland is highly fertile. Rice (*Oryza sativa* L.) is the main cereal cultivated today. The pulses cultivated include *Phaseolus vulgaris*, *P. aureus* Roxb., *P. mungo* L., and *Pisum sativum* L. Cash crops like *Brassica compestris* L. and to a lesser extent *Linum usitatissimum* L. are also grown. Orchards comprising the trees of *Malus sylvestris* Host, *Prunus* spp., and *Juglans regia* L. have been raised on the higher grounds. Also grown in the surrounding lowlands are *Populus* spp., *Salix* spp., *Ulmus* spp., *Robinia pseudoacacia* L., *Fraxinus excelsior* L., and other trees and shrubs.

The climate is similar to that of the rest of Kashmir, being essentially temperate. The seasons in the year are more or less well marked and can be placed under four distinct periods: winter (December to February), spring (March to May), summer (June to August), and autumn (September to November). The rainfall (fig. 2.2) is light and variable. The average rainfall is never more than a foot even in the month of January. The temperature (fig. 2.3) falls below freezing in winter and the highest temperature recorded in summer is never more than 35°C.

The site was first brought to light in 1935 by the Yale-Cambridge University Expedition Party of de Terra and Paterson, who laid some trial trenches which yielded a type of highly polished black ware and potsherds with unusual geometric designs assignable to a period ranging from 3,000

The Sites and their Environment 7

Figure 2.1: Burzahom: panoramic view.

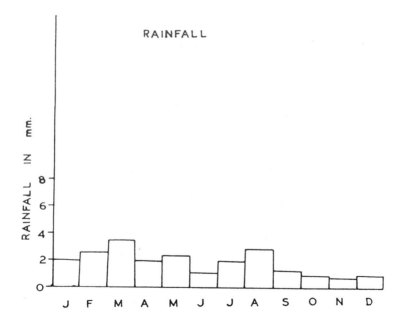

Figure 2.2: Burzahom: average monthly rainfall.

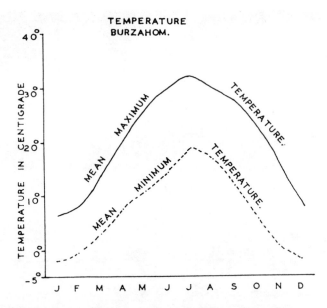

Figure 2.3: Burzahom: mean maximum and mean minimum temperature.

to 1,800 B.C. from the lower cultural layer. Subsequently, Archaeological Survey of India under the direction of Shri T.N. Khazanchi conducted systematic excavations at Burzahom for a series of seasons from 1961 to 1971.

The excavations have revealed that the site is primarily a Neolithic one capped by a Megalithic complex which consists of menhirs raised at a different stage. The excavations have brought to light four cultural periods (fig. 2.4).

Period I: Burzahom Neolithic I (2375–1700 B.C.)

The habitational pattern is distinguished by dwelling pits, circular or oval in plan, narrow at the top and wide at the base. There are also rectangular or squarish semi-subterranean shelters out about 0.50 m to 1 m into the natural soil, with post-holes, hearths, and drains. Pottery was one of the principal crafts and was mostly crude handmade ware, coarse in fabric and finish. The colours are chiefly steel grey, shades of dull red, brown, and buff. The pottery consists of bowls, vases, and stems. The people used tools of stone and bone. The tools were harpoons, needles with or without eyes, spear points, arrowheads, bores, and scrappers—all suggestive of hunting. Among the stone artefacts are axes, chisels, adzes, pounders, maceheads, points, picks, and hunting balls. There is no evidence of human burials.

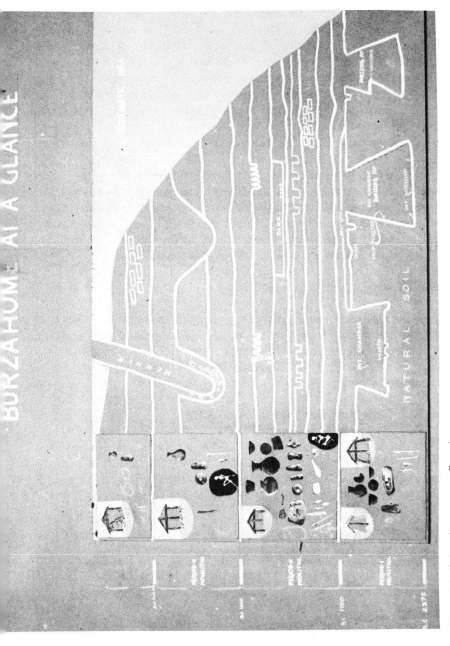

Figure 2.4: Cultural sequence at Burzahom.

Period II: Burzahom Neolithic II (1700–1000 B.C.)

The habitational pattern has undergone a change but the earlier cultural traits continue with some new cultural contacts. The pits are square or rectangular. There are numerous post-holes. The bulk of the pottery is handmade. A burnished black ware of medium fabric and the deluxe ware are two of the major wares. This period has yielded a great variety of pottery which is fine and the distinguished type is a high-necked jar in grey or black burnished ware. Other shapes are bowls, globular pots, jars, stems, and a funnel-shaped vessel. A few red ware, wheel-turned sherds and a globular pot painted in black and wheel turned have been found from the earliest levels.

The artefacts of stone and bone of this period are similar to the earlier phase but they are greater in number and better finished. A few copper-barbed arrowheads with the middle rib have also been found. Rectangular harvesters with a carved cutting edge and two holes on either side, double-edged picks in stone, and long bone needles with or without an eye have been reported. Human and animal burials have been found.

Period III: Megalithic Period (1000–600 B.C.)

This phase has been characterised by a number of megaliths. The most distinctive feature of this period is a rubble wall associated with megalith. The pottery has predominantly been red or gritty red ware. The few isolated copper tools and copper wire found during the Megalithic period seem to be a foreign intrusion.

Period IV: Early Historical (600 B.C.–200 A.D.)

The pit dwelling is replaced by the mud-brick structures. Though bone and stone tools are still used, their incidence is less. The pottery is predominantly wheel-turned red ware. Some metallic objects have also been retrieved, which point to the presence of iron in this period.

2.2. SEMTHAN

Semthan (75°9′ E longitude, 33°48′ N latitude) in Anantnag district is situated 44 km almost south of Srinagar on the Jammu-Srinagar National Highway, about 2 km from tehsil headquarters of Bijbehara (fig. 2.5). The site is located on the ancient lake bed of the valley floor and the serpentine Vitsta (Veth in Kashmiri) closely and calmly flows past. It comprises six low and high mounds: Tshradakut, Kuta, Rajmateng, Chakdhar, Guda, and Sonakhuta. The overall perimeter of ancient Semthan is 1.5 km to 2 km. At places the mounds rise to a height of 60 m above the surrounding ground level.

Figure 2.5: Semthan: panoramic view.

The site has been excavated by the Archaeological Survey of India under the direction of R.S. Bisht. The cultural sequence revealed the occurrence of the following occupational phases (fig. 2.6).

Period I: Pre-N.B.P. Phase (1500–600 B.C.)

The pre-N.B.P. phase is represented by a deposit of yellowish-brown, compact, sticky clay, distinguished by a type of ceramics most of which shows strong but remote genetic relationship with the late phase of the post-Harappan pottery of the Banawali-Bara phase of the plains of Punjab and Haryana, where it might be dated to 1500–1100 B.C. (Bisht 1977). The phase has revealed a few sherds of thick gritty red ware, a few of thick grey burnished ware, a sizable quantity of Bara-type pottery and a red ware that has apparently something to do with the Chalcolithic culture (Gaur 1987).

The site was not totally deserted by the inhabitants of this period before the arrival of their successors, since there was a cultural overlap between the two, evidenced by the continuation of the early in the later and thereby a sort of fusion at a stage.

12   *Palaeoethnobotany*

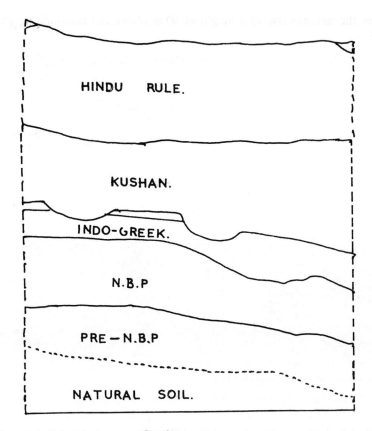

Figure 2.6: Cultural sequence at Semthan.

Period II: N.B.P. Phase (600–200 B.C.)

The inhabitants of the N.B.P. had a distinct material culture, including a remarkable class of pottery known as the N.B.P. (northern black polished ware) along with the typical associated grey and red ware. A noticeable feature in the construction is the use of mud clods, especially in making floors. A rubble of stone wall was also revealed. The use of iron is evident. Beads of terracotta, semi-precious stone, and crystalline quartz were also recovered.

Period III: Indo-Greek Phase (200–1 B.C.)

Sandwiched between N.B.P. and Kushan horizons, a comparatively thin deposit consisting of various mud floors and horizontally running streaks yielded a very distinct class of pottery which is reddish-pink slipped and thin in fabric. It seems to be associated with the Indo-Greeks. This phase had never been found in the valley earlier.

Period IV: Kushan Phase (1–500 A.D.)

The Kushan phase is represented by a deposit of brownish compact earth and characterized by wheel-turned red ware. Some burnt bricks have been met with. Floor levels paved with stones were also encountered.

Period V: Hindu Rule Phase (500–1000 A.D.)

The pottery of the Hindu Rule phase is wheel turned with fine-grained fabric and lustrous red slip. This phase corresponds to a period when temple building activity was achieved by the Hindu rulers in the valley.

CHAPTER 3

KEYS AND CRITERIA FOR IDENTIFICATION

As has been indicated in the introductory chapter, the preliminary investigations provided a fair impression of the diversity of plant species and groups in the archaeological remains. Accordingly, studies were directed toward developing the keys and criteria for identification which are summarized in the following pages.

3.1. CEREALS

3.1.1. Wheats (*Triticum* Spp.)

A grain of wheat, often spoken of as a 'seed' by farmers, is a single-seeded kernel known as 'caryopsis', the name first used by Richard (1819). The grain is more or less ovoid with a tuft of hairs known as 'brush' at the apex. A groove or crease extends from end to end of the grain on the ventral side and on the opposite (dorsal) side the outline of the embryo can be seen at the base of the grain.

Wheat being one of the most important present-day food plants, extensive and arduous research has been carried out on it from different points of view. Based on the number of chromosomes that various species possess, the genus is grouped under diploid ($2n = 14$), tetraploid ($2n = 28$) and hexaploid ($2n = 42$) species. The genus has also been classified on the basis of morphological characters. Those species in which the kernel is not released during threshing and has glumes attached to it at maturity are called 'glume wheats' and those in which the glume is not present at maturity are called 'naked wheats'. Naked wheats include those species which have their kernel loose at maturity and detach from the chaff easily (Helbaek 1960a, Paterson 1965). Classification of the genus as provided by Zohary (1971) is presented in table 3.1.

Table 3.1. Wheats: Main morphological types based on cytogenetic affinities.

Diploid einkom wheats, 2n = 14 (genome A) Collective name: *T. monococcum* L	1. *T. boeoticum* Boiss emend Schiem (wild, brittle, hulled) 2. *T. monococcum* L (cultivated, hulled)
Tetraploid emmers, durums and turgidums, 2n = 28 (genomes AB) Collective name: *T. turgidum* L	1. *T. dicoccoides* (Kornicke) Aarons (wild, brittle, hulled) 2. *T. dicoccum* Schubl (cultivated, hulled) 3. *T. durum* Desf (cultivated, free threshing) 4. *T. turgidum* L (cultivated, free threshing) 5. *T. persicum* Vav (cultivated, free threshing) 6. *T. polonicum* L (cultivated, free threshing)
Tetraploid, timopheevii wheats 2n = 28 (genomes AG) Collective name: *T. timopheevii* Zhuk	1. *T. araraticum* Jakubz (wild, brittle, hulled) 2. *T. timopheevii* Zhuk (cultivated, hulled)
Hexaploid wheats, 2n = 42 (genomes ABD) Collective name: *T. aestivum* L	1. *T. spelta* L (cultivated, hulled) 2. *T. macha* Dekr et Men (cultivated, hulled) 3. *T. aestivum* L (cultivated, free threshing) 4. *T. compactum* Host (cultivated, free threshing) 5. *T. sphaerococcum* Perc (cultivated, free threshing) 6. *T. vavilovii* Jakubz (cultivated, free threshing)

The mode of fracturing of rachis in glume wheats is also used in identifying the species (Watkins 1930, Helbaek 1964a, Peterson 1965). In domesticated forms of glume wheats the rachis breaks into spikelets when threshed. On the other hand, in wild glume wheats, the rachis breaks on ripening while still on the erect plant and disperses the fruits, thus making it impossible to collect the grains (fig. 3.1). In naked wheats, the rachis being non-fracturing, it is seldom found with the kernel. For instance, in *T. boeoticum* and *T. dicoccoides* (both wild) and *T. monococcum* and *T. dicoccum* (both cultivated) the rachis fractures at joints in such a way that a segment of rachis is left attached below each spikelet. However, in *T. spelta*, the rachis does not fracture at the joints but breaks at another point in such

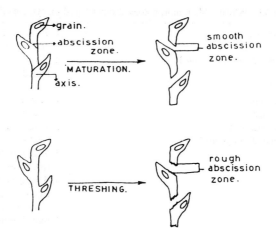

Figure 3.1: Mode of fracturing in wild (above) and domesticated (below) forms of wheat inflorescence.

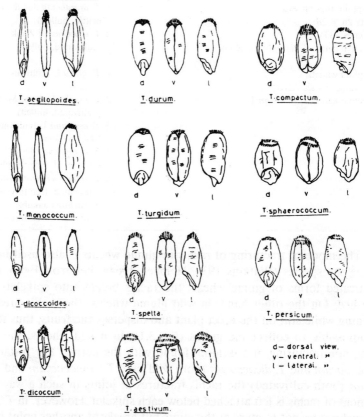

Figure 3.2: Exomorphology of *Triticum* spp.

a way that a piece of rachis remains attached to the face of the spikelet directed upwards.

EXOMORPHIC CHARACTERS (FIG. 3.2)

The colour, shape and size of caryopses of various species of *Triticum* vary a great deal. The colour of caryopses is olive yellow in *T. aegilopoides* syn. *T. boeoticum*, yellowish in *T. monococcum, T. dicoccum, T. durum, T. turgidum, T. compactum*, yellowish-brown in *T. persicum* and *T. aestivum*, and reddish-brown in *T. sphaerococcum, T. spelta*, and *T. dicoccoides*.

Grains are oval in *T. compactum, T. aestivum*, and *T. turgidum*, elliptic in *T. spelta, T. persicum, T. durum, T. dicoccoides*, and *T. aegilopoides*, oval with tapering ends in *T. monococcum* and *T. dicoccum*, and spherical in *T. sphaerococcum*.

Average length varies from 4.8 mm in *T. sphaerococcum* to 9 mm in *T. dicoccoides*. Similarly, the average breadth varies from 1.7 mm in *T. dicoccoides* to 3.3 mm in *T. spelta* and *T. compactum*. The various species show increase in breadth and decrease in length as one proceeds from diploid and tetraploid to hexaploid species.

ENDOMORPHIC CHARACTERS (FIGS. 3.3–3.14)

Wheat grain has six main parts, from interior outwards, embryo, starchy endosperm, aleurone layer, nucellus, seed coat, and pericarp. The first five of these constitute the seed; pericarp is not included as a part of the seed because it is derived from the ovary wall. The pericarp and seed coat are fused together. Cell pattern and cell alignment of various layers of pericarp have proven taxonomically useful.

The outer epidermis is made up of rectangular parenchymatous cells in *T. aegilopoides, T. monococcum, T. dicoccum, T. dicoccoides, T. turgidum, T. persicum, T. compactum, T. sphaerococcum*, and *T. aestivum*, with some polygonal cells in *T. spelta* and trapezium-shaped cells in *T. durum*.

The cell wall is wavy in *T. aegilopoides*, smooth in *T. dicoccoides, T. dicoccum*, and *T. spelta*, septate in *T. persicum* and *T. compactum*, and pitted in *T. monococcum, T. durum, T. turgidum, T. sphaerococcum*, and *T. aestivum*.

The partition wall between the two adjacent cells is oblique in *T. aegilopoides, T. dicoccoides, T. dicoccum*, and *T. durum*, perpendicular in *T. monococcum, T. aestivum*, and *T. turgidum*, and both oblique and perpendicular in *T. spelta, T. persicum, T. compactum*, and *T. sphaerococcum*. The epidermal cells lie parallel to the grain axis in all the species.

Next to the epidermal layer is the 'cross layer' made up of hexagonal cells in *T. dicoccoides* and rectangular cells in *T. aegilopoides, T. turgidum, T. compactum*, and *T. sphaerococcum*. Both hexagonal and pentagonal shapes

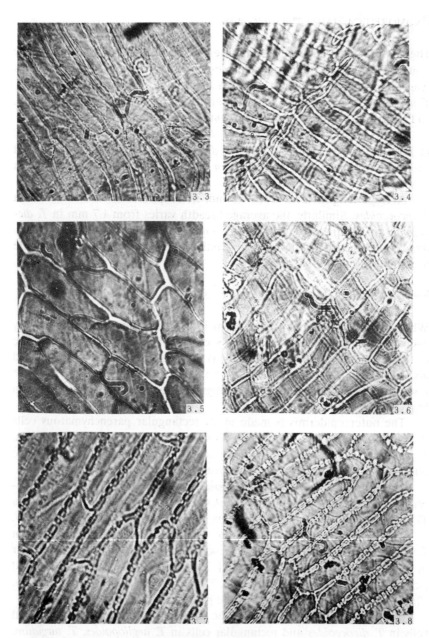

Figure 3.3: Outer epidermis of *T. aegilopoides* 400x.
Figure 3.4: Cross layer of *T. aegilopoides* 400x.
Figure 3.5: Outer epidermis of *T. dicoccum* 400x.
Figure 3.6: Cross layer of *T. dicoccoides* 400x.
Figure 3.7: Outer epidermis of *T. sphaerococcum* 400x.
Figure 3.8: Cross layer of *T. durum* 400x.

Keys and Criteria for Identification

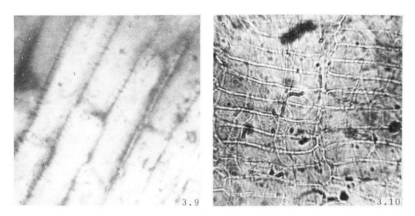

Figure 3.9: Outer epidermis of *T. persicum* 400x.   Figure 3.10: Cross layer of *T. spelta* 400x.

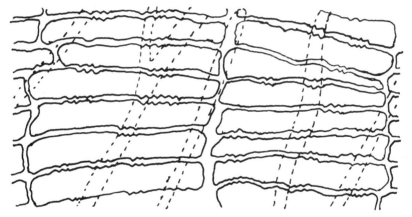

Figure 3.11: Camera lucida drawing showing wavy cell wall in *T. aegilopoides* 400x.

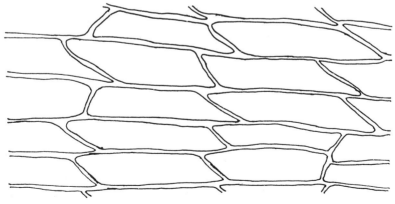

Figure 3.12: Smooth cell wall in *T. dicoccoides* 400x.

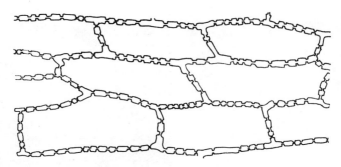

Figure 3.13: Beaded cell wall in *T. sphaerococcum* 400x.

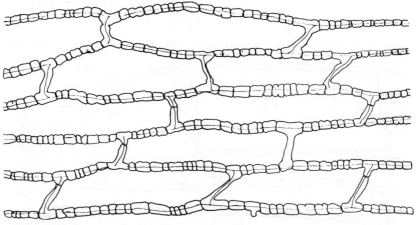

Figure 3.14: Septate cell wall in *T. compactum* 400x.

have been observed in *T. dicoccum* and *T. aestivum*, rectangular to pentagonal in *T. spelta* and *T. durum*, and polygonal in *T. monococcum* and *T. persicum*. The cells are arranged at right angles to the cells of the outer epidermal layer, thus across the grain in all the species.

The inner epidermis is made up of thin-walled tube cells in all the species.

The aleurone layer is composed of isodiametric cells in *T. aegilopoides, T. dicoccum, T. durum, T. persicum, T. compactum, T. aestivum*, polygonal in *T. monococcum*, rectangular to polygonal in *T. dicoccoides*, and polygonal to isodiametric in *T. turgidum, T. sphaerococcum*, and *T. spelta*.

SCANNING ELECTRON MICROGRAPH (FIGS. 3.15–3.25)

In the scanning electron micrograph (SEM) the pericarp shows differences in cell alignment, shape, and wall thickness, forming distinct cell patterns and relief in hulled as well as naked wheats. Among hulled wheats *T. aegilopoides* and *T. dicoccoides* are conspicuous by the shape of outer epidermal cells. The end walls of these cells are obliquely placed in *T. aegilopoides*, and *T. dicoccoides*. The two wild species can be distinguished from their cultivated counterparts *T. monococcum* and *T. dicoccum* respectively. The three cultivated hulled species, *T. monococcum, T. dicoccum*, and *T. spelta*, cannot be distinguished easily from their surface features. On the other hand, naked wheats *T. durum, T. turgidum, T. aestivum, T. compactum, T. persicum*, and *T. sphaerococcum* can easily be distinguished from each other with the help of SEM features.

The results of various exo-endomorphic features of caryopses are presented in table 3.2. It has been possible to prepare a key for the identification of various species of *Triticum* based on the structure of pericarp as follows:

KEY TO THE SPECIES OF *TRITICUM*

1	+	Cells of outer epidermal layer with smooth or wavy walls.	... 2
	−	Cells of outer epidermal layer with perforated or septate walls.	... 5
2	+	Cells with smooth walls.	... 3
	−	Cells of outer epidermal layer and cross layer with wavy walls.	*T. aegilopoides*
3	+	Cells of epidermal layer and cross layer with smooth walls.	... 4
	−	Cells of epidermal layer with smooth walls and of cross layers with septate walls.	*T. spelta*
4	+	Cells of outer epidermal layer comparatively long (117–140 μm) and of cross layer comparatively short (162–182.5 μm). Aleurone layer with rectangular and polygonal cells.	*T. dicoccoides*
	−	Cells of outer layer comparatively short (62–165 μm) and of cross layer comparatively long (230–247 μm). Aleurone layer with isodiametric cells.	*T. dicoccum*
5	+	Cells of outer epidermal layer septate.	... 6
	−	Cells of outer epidermal layer perforated.	... 7
6	+	Cells of cross layer with smooth walls.	*T. compactum*
	−	Cells of cross layer with perforated walls.	*T. persicum*

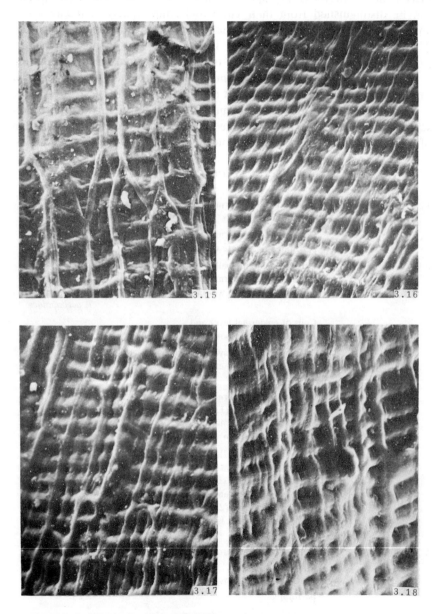

Figure 3.15: SEM of *T. aegilopoides* 200x.
Figure 3.16: SEM of *T. monococcum* 200x.
Figure 3.17: SEM of *T. dicoccoides* 200x.
Figure 3.18: SEM of *T. dicoccum* 200x.

Keys and Criteria for Identification

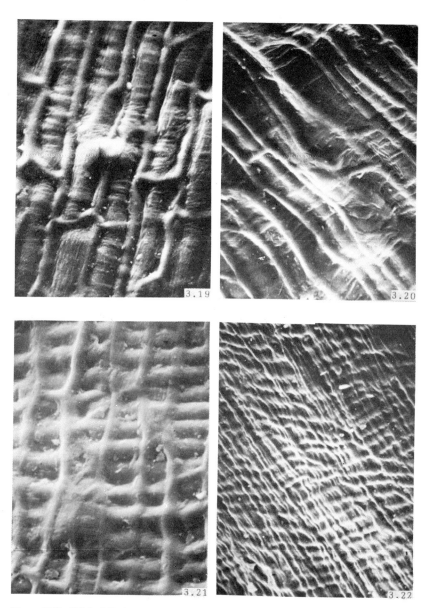

Figure 3.19: SEM of *T. durum* 200x.
Figure 3.20: SEM of *T. turgidum* 200x.
Figure 3.21: SEM of *T. persicum* 200x.
Figure 3.22: SEM of *T. spelta* 200x.

Table 3.2 Exo- and endomorphic features of

Species	Exomorphology		Endomorphology					
	Shape	Average length x average breadth (mm)	Outer epidermal layer					
			Cell shape	Cell wall	No. of pits/cell	Partition wall	Average cell length (μm)	Average cell breadth (μm)
Triticum aegilopoides	Oval, pointed at both ends.	6.7 x 2.5	Rectangular	Wavy	—	Oblique	145	26.8
T. monococcum	Oval, tapering at both ends.	7.5 x 2.1	Rectangular	Pitted	10–12	Perpendicular	111	14.5
T. dicoccoides	Narrow, pointed at both ends.	9 x 1.7	Rectangular	Smooth	—	Oblique	136.5	28.2
T. dicoccum	Narrow, pointed at both ends.	7.3 x 2.8	Rectangular	Smooth	—	Oblique	121.5	35.1
T. durum	Oval	7.1 x 3.1	Rectangular and trapezium shaped	Pitted	15–24	Oblique	274.3	31.6
T. turgidum	Oval, broad, plump	6.4 x 3.1	Rectangular	Pitted	16–26	Perpendicular	179.5	37
T. persicum	Narrow, pointed at both ends	8.5 x 1.9	Rectangular	Septate	28–30 septa/cell	Pitted, perpendicular and oblique	217	55.2

pericarp of caryopses of *Triticum* spp.

	Endomorphology						
	Cross layer				Aleurone layer		
Cell shape	Cell wall	No. of pits/ cell	Average cell length (μ m)	Average cell breadth (μ m)	Cell shape	Average diameter (μm)	Remarks
Rectangular	Wavy	—	138	15.5	Isodiametric	49	Wild (diploid)
Polygonal	Pitted	10–13	193.3	23.3	Polygonal with rounded corners	63	Glume wheat (diploid)
Hexagonal	Smooth	—	183	14.5	Rectangular and polygonal	45	Wild (tetraploid)
Pentagonal and hexagonal	Smooth	—	236	24	Isodiametric	41	Glume wheat (tetraploid)
Rectangular and pentagonal	Pitted	15–18	194	24.5	Isodiametric	65	Naked wheat (tetraploid)
Rectangular	Pitted	18–22	197	17.5	Isodiametric pentagonal and hexagonal	57.5	Naked wheat (tetraploid)
Rectangular and hexagonal	Pitted	25–32	167.5	18.7	Isodiametric	63	Naked wheat (tetraploid)

Table 3.2. (contd.)

T. compactum	Oval, plump and narrow towards apex	6.2 × 3.3	Rectangular	Septate	25–30 Septa/cell	Pitted, perpendicular and oblique	237	57.5
T. sphaerococcum	Rounded and angular	4.8 × 3	Rectangular	Pitted	24–31	Perpendicular and oblique	167.9	54.1
T. spelta	Pointed at both ends	8.5 × 3.3	Rectangular and pentagonal	Smooth	—	Perpendicular and oblique	283.5	73.5
T. aestivum (T. vulgare)	Oval, plump blunt at apex	6.3 × 3.0	Rectangular	Pitted	14–28	Pitted and perpendicular	250.3	30

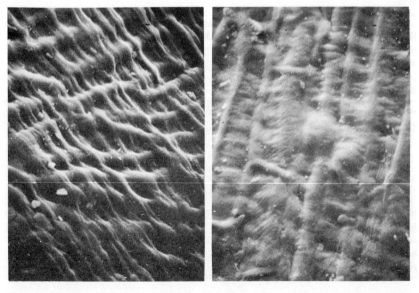

Figure 3.23: SEM of *T. aestivum* 200x.   Figure 3.24: SEM of *T. compactum* 200x.

Keys and Criteria for Identification 27

Rectangular	Smooth	—	297	33.5	Isodiametric	43	Naked wheat (hexaploid)
Rectangular	Septate	18–42 Septa/cell	191.4	14.5	Isodiametric and polygonal	59	Naked wheat (tetraploid)
Rectangular and Pentagonal	Septate	23–42 Septa/cell	147.5	21.7	Isodiametric and polygonal	73	Glume wheat (hexaploid)
Rectangular and hexagonal	Pitted	16–25	260.3	16.5	Isodiametric	48.5	Naked wheat (hexaploid)

Figure 3.25: SEM of *T. sphaerococcum* 200x.

7	+	Cells of cross layer with perforated walls.		... 8
	−	Cells of cross layer with septate walls.		*T. sphaerococcum*
8	+	Partition walls between two adjacent cells in outer epidermal layer perpendicular.		... 9
	−	Partition wall in outer epidermal layer obliquely placed; cells rectangular to trapezium-shaped.		*T. durum*
9	+	Number of pits per cell in outer epidermal layer 14–28.		... 10
	−	Number of pits per cell in outer epidermal layer 10–12.		*T. monococcum*
10	+	Cells of outer epidermal layer elongate, rectangular; 110–267.5 μm long; 25–40 μm broad; 16–26 pits per cell. Cells of cross layer elongate, 160–212.5 μm long and 7.5–22.5 μm broad; 18–22 pits per cell.		*T. turgidum*
	−	Cells of outer epidermal layer elongate, rectangular; 140–372.5 μm long; 22.5–37.5 μm broad; 14–28 pits per cell. Cells of cross layer elongate, pentagonal; 215–390 μm long; 12.5–30 μm broad; 16–25 pits per cell.		*T. aestivum*

3.1.2. Barley (*Hordeum* Spp.) (fig. 3.26)

The colour, shape, and size of caryopses vary to a great extent in different species of *Hordeum*. The colour of the caryopses varies from yellow in *H. spontaneum*, *H. vulgare* var. *nudum*, and *H. hexaploidum* to reddish-yellow in *H. distichum* and *H. vulgare* var. *hulled*. Shape of caryopses ranges from elliptic in *H. spontaneum* and *H. hexaploidum* to oval in *H. distichum* and *H. vulgare*. The grains are symmetrical in *H. vulgare* var. *nudum* and asymmetrical in *H. vulgare* var. *hulled*, *H. hexaploidum*, *H. spontaneum*, and *H. distichum*. The average size of various species of *Hordeum* ranges from 7.3 mm × 3.2 mm in *H. vulgare* var. *nudum* to 14.1 mm × 2.9 mm in *H. spontaneum*.

ENDOMORPHIC CHARACTERS (FIGS. 3.27–3.33)

The grain is composed of six main parts. Proceeding from interior outwards, these are the embryo, starchy endosperm, aleurone layer, nucellus, seed coat, and pericarp. The first five of these constitute the seed.

In hulled species, after the husk (lemma and palea) is removed, the outer epidermal layer is made up of sinuous-walled rectangular cells with oval to round hair bases and rounded to elliptical silica bodies in *H. spontaneum, H. distichum, H. hexaploidum*, and *H. vulgare* var. *hulled*, and elongate rectangular, smooth-walled cells without hair bases and silica bodies in *H. vulgare* var. *nudum*. The cells lie parallel to the grain axis in all species.

Next to the epidermal layer is the cross layer, consisting of rectangular cells in *H. spontaneum, H. distichum, H. vulgare* var. *hulled*, and *H. vulgare* var. *nudum*, and squarish to hexagonal cells in *H. hexaploidum*. The cells are arranged at right angles to the epidermal cells in *H. hexaploidum* and *H. spontaneum*. In *H. distichum* and *H. vulgare* both *hulled* and *nudum*, cells lie parallel to the grain axis. The cell wall is smooth in *H. hexaploidum, H. vulgare* both *hulled* and *nudum*, and *H. spontaneum* and sinuous with stomata in *H. distichum*.

The inner epidermal layer consists of rectangular to squarish cells in *H. spontaneum* and rectangular cells in *H. hexaploidum, H. distichum*, and *H. vulgare* both *hulled* and *nudum*. The cells lie parallel to the grain axis.

The aleurone layer consists of squarish to rectangular cells in *H. spontaneum, H. vulgare* var. *nudum*, and *H. hexaploidum*, polygonal to hexagonal cells in *H. vulgare* var. *hulled*, and rectangular to rounded cells in *H. distichum*.

SEM (FIGS. 3.34–3.37)

In SEM *Hordeum spontaneum* shows abundant silica bodies, *H. vulgare* var. *hulled* shows sinuous-walled long cells and rounded to spherical short cells with silica bodies, *H. vulgare* var. *nudum* shows elongate, rectangular, smooth-walled cells with perpendicular end walls, and *H. distichum* shows elongated sinuous-walled long cells with some rounded to spherical short cells. Thus, various species can be distinguished on the basis of SEM study.

Morpho-anatomical features of various species of *Hordeum* are given in table 3.3. It has been possible to prepare a key for their identification on the basis of caryopsis structure as follows:

KEY TO THE SPECIES OF *HORDEUM*

1	+	Cells of the outer epidermal layer with sinuous walls.	... 2
	−	Cells of the outer epidermal layer with smooth walls.	*H. vulgare* var. *nudum*
2	+	Cells of outer epidermal layer sinuous-walled with wide lumen.	... 3
	−	Cells of outer epidermal layer	

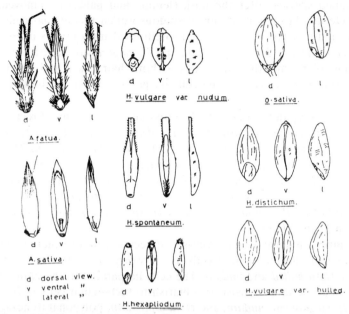

Figure 3.26: Exomorphology of *Hordeum*, *Avena* and *Oryza* spp.

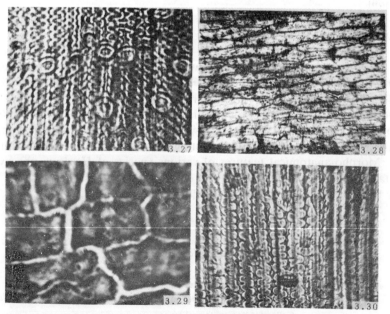

Figure 3.27: Outer epidermis of *Hordeum spontaneum* 400x.
Figure 3.28: Cross layer of *H. spontaneum* 400x.
Figure 3.29: Inner epidermis of *H. spontaneum* 400x.
Figure 3.30: Outer epidermis of *H. distichum* 400x.

Keys and Criteria for Identification 31

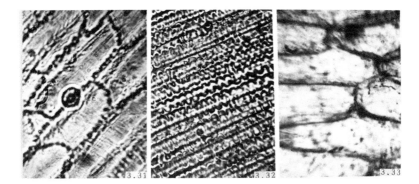

Figure 3.31: Cross layer of *H. distichum* 400x.
Figure 3.32: Outer epidermis of *H. vulgare* hulled 400x.
Figure 3.33: Cross layer of *H. hexaploidum* 400x.

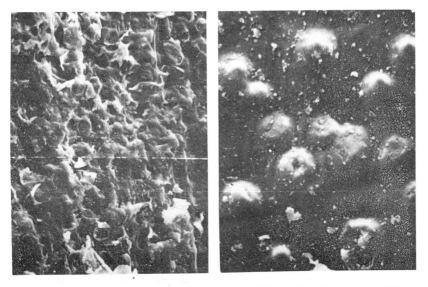

Figure 3.34: SEM of *H. distichum* 200x.   Figure 3.35: SEM of *H. spontaneum* 200x.

		sinuous-walled with narrow lumen, 8–32 projections per cell, cells of cross layer elongate, 175–565.5 μm long, occasionally pointed at one end. Partition wall between cells swollen.	*H. vulgare* var. *hulled*
3	+	Hair bases oval to round; cells of cross layer parallel to epidermal cells and rectangular to columnar in shape.	... 4
	−	Hair bases with wavy margins; cells of outer epidermal layer having 12–36 projections per cell; cells of cross layer at right angles to outer epidermal cells and polygonal with smooth and thick walls.	*H. hexastichum*
4	+	Cells of outer epidermal layer thin-walled with 14–20 projections per cell; cells of cross layer with sinuous walls having 2–6 projections per cell and provided with stomata.	*H. distichum*
	−	Cells of outer epidermal layer thick-walled with 16–36 projections per cell; cells of cross layer smooth-walled, elongate, columnar and occasionally pointed at one end.	*H. spontaneum*

3.1.3. Oats (*Avena* Spp.)

EXOMORPHIC CHARACTERS (FIG. 3.26)

Avena spp. are classified into three groups based on the number of chromosomes, as in table 3.4. Two species, *Avena fatua* and *A. sativa*, are found in this region.

The grains of *Avena* spp. are elongate with pointed base, blunt apex, and average size 9.9 mm × 2 mm in *A. sativa* and pointed at both ends with average size 8 mm × 1.5 mm in *A. fatua*.

ENDOMORPHIC CHARACTERS (FIGS. 3.38–3.41)

The outer epidermal layer is made up of elongate rectangular cells with some cells pointed at one end in *A. fatua* and elongate rectangular cells in *A. sativa*. Spherical to rounded hair bases are present in *A. fatua* and absent in *A. sativa*. Cells are arranged parallel to the axis of grain.

Keys and Criteria for Identification

Table 3.4. *Avena*: Classification into species (after Martin and Leonard 1967).

Diploid group (2n = 14)	Tetraploid group (2n = 28)	Hexaploid group (2n = 42)
Avena brevis Roth (short oat)	*Avena barbata* Brot (slender oat)	*Avena fatua* Linn (common wild oat)
A. wiestii Steud (desert oat)	*A. abyssinica* Hochst (Abyssinian oat)	*A. sativa* Linn (white oat)
A. strigosa Schreb (sand oat)		*A. nuda* Linn (large-seeded naked oat).
A. nudibrevis Vav (small-seeded naked oat).		*A. sterilis* Linn (wild, red oat)
		A. byzantina Koch (cultivated, red oat).

The cross layer consists of elongate, rectangular to squarish cells lying at right angles to epidermal cells. The cell wall is wavy in *A. fatua* and smooth in *A. sativa*.

The aleurone layer is made up of spherical cells in both *A. sativa* and *A. fatua*.

SEM (FIGS. 3.42–3.43)

In SEM the oats show rectangular, elongate, smooth-walled cells. However, the cells of *Avena fatua* are smaller than those of *A. sativa*.

Morpho-anatomical features of two species of *Avena* are given in table 3.5. The two species can be identified on the basis of pericarp structure as follows:

KEY TO THE SPECIES OF *AVENA*

| 1 | + | Outer epidermal cells columnar with one end pointed; partition wall between two cells oblique, rounded hair bases present; cells of cross layer squarish to rectangular with wavy wall. | *A. fatua* |
| | – | Outer epidermal cells rectangular; partition wall between two cells horizontal, cells of cross layer squarish with smooth wall. | *A. sativa* |

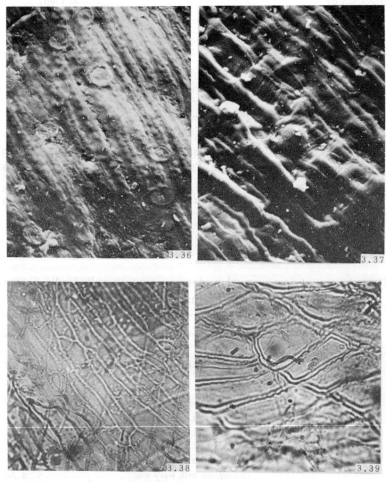

Figure 3.36: SEM of *H. vulgare* hulled 200x.
Figure 3.37: SEM of *H. vulgare* var. *nudum* 200x.
Figure 3.38: Outer epidermis of *Avena fatua* 400x.
Figure 3.39: Cross layer of *A. fatua* 400x.

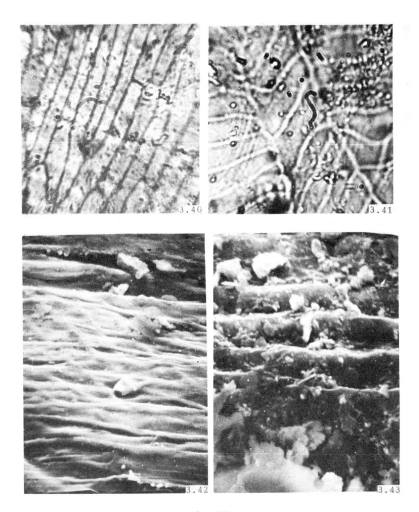

Figure 3.40: Outer epidermis of *A. sativa* 400x.
Figure 3.41: Cross layer of *A. sativa* 400x.
Figure 3.42: SEM of *A. fatua* 200x.
Figure 3.43: SEM of *A. sativa* 200x.

Table 3.5. Exo- and endomorphic features of

Species	Exomorphology		Endomorphology			
			Outer epidermal layer			
	Shape	Average length x average breadth (mm)	Cell shape	Cell wall	Average cell length (μm)	Average cell breadth (μm)
Avena fatua	Elongate, pointed at both ends	8 x 1.5	Rectangular, some cells having one end pointed	Smooth	353.5	29
A. sativa	Elongate, with pointed base and blunt apex	9.9 x 2	Rectangular, some cells having one end rounded	Smooth	118.5	33

3.1.4. Rice (*Oryza sativa*) (figs. 3.26, 3.44–3.45)

The caryopses of *Oryza sativa* are yellow, ellipsoidal, and flattish on both sides with pointed base and blunt apex. The caryopses are covered with hairs which are dense at ridges. Two bracts or glumes are present at the base. At the apex it is V-shaped, and one arm of the V is elongated to form an awn, which is hairy. Caryopses bear two ridges on one side and the embryo is lateral. Caryopses have an average length of 7.7 mm and average breadth of 1.9 mm.

The epidermal layer consists of elongate rectangular cells, 87 μm to 112 μm long and 40 μm to 125 μm broad, placed lengthwise from base to apex of the grain, thus parallel to grain axis. The lumen is narrow. The cell wall is thrown up into narrow, acute pointed and straight projections spread out into the entire space of the lumen of the cells and overtailing with the projections of the opposite cell. Cell wall is thicker than the lumen of the cells, varying in thickness from 45 μm to 70 μm; 6 to 8 projections are present per cell.

Hair bases are rounded structures present frequently at the base of the epidermal cells. SEM of husk shows a characteristic chessboard pattern.

3.2. MILLETS

Morpho-anatomical features of millet species *Eleusine coracana, Panicum miliaceum, Paspalum sacrobiculatum, Setaria glauca, S. italica, S. viridis*, and *Sorghum bicolor* are provided in table 3.6.

pericarp of caryopses of *Avena* spp.

Endomorphology						
Inner epidermal layer				Aleurone layer		
Cell shape	Cell wall	Average cell length (μm)	Average cell breadth (μm)	Cell shape	Average cell diameter (μm)	Remarks
Rectangular, some cells shorter than broad	Smooth	85.5	63	Spherical to rounded	71.5	Wild oat (hexaploid)
Rectangular and squarish	Smooth	153	59.5	Spherical	42.5	White oat (hexaploid)

3.3. PULSES

Corner (1951) has emphasized that a seed coat having an outer palisade and hour glass cells below it is apparently identifiable as leguminous. Chowdhury and Buth (1970) have made a comprehensive study of the seed coat anatomy of Indian pulses and provided a key based on size, shape, and surface of seeds; size and position of hilum; shape and height of palisade cells; and the characteristic structure of the cuticle (table 3.7). This key is very useful in determining the identity of archaeological pulses. Hopf (1986) has also dealt with the seed coat structure and its utility in identification of pulse species.

3.4. HORTICULTURAL FRUITS

The evidence of horticultural crops in the archaeological plant material usually consists of remains of 'stone fruits'. The pericarp of the 'stone fruits' or 'drupes' consists of an outer epicarp, a fleshy mesocarp, and the innermost stony and hard endocarp. The endocarp is the most likely part of the fruit to be archaeologically preserved. Therefore, in the present study endocarps of various species were investigated from the morpho-anatomical point of view.

EXOMORPHIC CHARACTERS

The colour, texture, shape, and size of endocarps of various species of *Prunus*, *Juglans*, and *Zizyphus* vary a great deal. In colour, endocarps are brown in *Juglans regia* and *Zizyphus jujuba*, yellowish-brown in *Prunus armeniaca* and *P. persica*, creamy in *P. domestica* and *P. cerasus*, and yellow in *P. amygdalus*. Endocarps vary in texture from thin and papery in *Prunus amygdalus* to very thick in *P. persica*. In shape, endocarps vary to a great extent: in *P. amygdalus* and *P. armeniaca* endocarps are ovoid to oval and slightly compressed on both sides; in *P. persica* ovoid and compressed; in *P. domestica* and *Zizyphus jujuba* elliptic to oval; in *P. cerasus* ovoid to spherical and raised on both sides; and in *J. regia* spherical. The outer surface is deeply sculptured in *P. persica*, pitted in *P. amygdalus*, grooved and irregularly pitted in *J. regia*, with raised projections in *Z. jujuba*, with ripple markings in *P. armeniaca*, and smooth in *P. domestica* and *P. cerasus*. The inner surface is minutely sculptured in *P. domestica*, with raised projections in *J. regia*, and smooth in *Z. jujuba*, *P. persica*, *P. cerasus*, *P. armeniaca* and *P. amygdalus*.

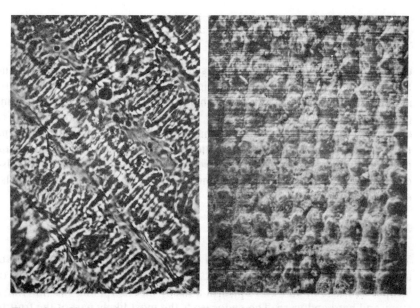

Figure 3.44: Husk of *Oryza sativa* 400x.

Figure 3.45: SEM of *O. sativa* show- ing chessboard pattern.

ENDOMORPHIC CHARACTERS (FIGS 3.46–3.52)

Endocarps are mainly composed of sclereids of various types. In the present study, after maceration four types of sclereid have been observed.

Table 3.6. Morpho-anatomical features of some species of millets.

Species	Shape and surface	Epidermis
Eleusine coracana	Circular, hulled, ventral surface longitudinally depressed, glumes absent	Smooth-walled, cells longer than broad
Panicum miliaceum	Oval, lemma and palea smooth or roughened by fine tubercles on dorsal convex side	Smooth or sinuous-walled, encrusted with silica bodies, lumen narrow
Paspalum sacrobiculatum	Oval, convex on dorsal side, tightly enclosed within lemma and palea, tubercles on dorsal side	Slightly sinuous, broader than long cells
Setaria glauca	Oval, convex on the dorsal side, tightly enclosed within lemma and palea, tubercles transversely oriented on dorsal surface	Deeply sinuous cells
Setaria italica	Longer than broad, oval, glumed, lemma flattened, palea convex, tubercles on dorsal side	Dentate, longer than broad cells
Setaria viridis	Flattened ventrally, pointed at the apex, obtuse base, glossy dorsal surface	Deeply sinuous, irregular cells
Sorghum bicolor	Rounded, bluntly pointed obovate, obtuse base, glumed, without tubercles	Smooth, longer than broad cells

1. Macrosclereids: elongate, columnar.
2. Brachysclereids: small, rounded, somewhat isodiametric.
3. Asterosclereids: star-shaped.
4. Osteosclereids: elongate, bone-shaped.

In *Prunus amygdalus, P. domestica*, and *Zizyphus jujuba* endocarps revealed a majority of macrosclereids with a few brachysclereids; in *P. persica* it is formed mainly of macrosclereids with some osteosclereids and brachysclereids; in *P. armeniaca* it is formed of a majority of brachysclereids with some macrosclereids; in *Juglans regia* it is formed of a majority of asterosclereids with a few brachysclereids; and in *P. cerasus* it is formed of macrosclereids only.

Morpho-anatomical characters of various species of *Prunus, Juglans*, and *Zizyphus* are given in table 3.8. From the above account it is quite evident that the endocarps of various genera can be easily separated with the help of morpho-anatomical features. A key for identification of various species of *Prunus* is as follows:

Anatomy of Indian Pulses

Hilum Position	Palisade cells		Cuticle
	Shape	Height	
Partially covered with whitish hard tissue raised above the level of seed surface	Type I	89 ± 13	Smooth, thin
In a sunken pouch below the level of seed surface	Type III	137 ± 27	Slightly rough
Below the level of seed surface partially covered with whitish tissue	Type II	94 ± 5	Smooth
Completely covered with whitish hard tissue	Type I	64 ± 10	Smooth
Above the level of seed surface along prominent shape running many mm	Type II	134 ± 8	Smooth
In level with seed surface whitish hard tissue absent	Type II	96 ± 8	Dentate
Almost in level with seed surface whitish hard tissue absent	Type I	47±5	Dentate
Completely covered with whitish hard tissue	Type I	56 ± 9	Smooth
Completely covered with whitish hard tissue	Type I	56±7	With papillae or papillae-like outgrowth

Table 3.7.

Phaseolus mungo L.	5.1 × 4.2	Smooth	3 × 2
Pisum sativum L.	8.6 × 6.6	Smooth	3 × 3
Vicia faba L.	8.4 × 7	Smooth	4.6 × 2
Vigna catuang Walp.	8.8 × 5.4	Smooth	4.6 × 2
Vigna sinensis Savi ex Hassak.	14.3 × 7.5	Smooth	3 × 2

Note: Type 1 ▯ Type 2 ▯ Type 3 ▯

KEY TO THE SPECIES OF *PRUNUS*

1	+	Outer surface of endocarp smooth.	... 2
	−	Outer surface of endocarp pitted or scluptured.	... 3
2	+	Inner surface minutely scluptured; endocarp elliptic to oval, 7–12 mm in size, having macrosclereids in majority with some brachysclereids.	*P. domestica*
	−	Inner surface smooth, endocarp round, raised on both sides, 17–28 mm in size with only macrosclereids.	*P. cerasus*
3	+	Outer surface pitted, endocarp ovoid to oval, compressed on both sides, 25–40 mm in size, having majority of macrosclereids with some brachysclereids.	*P. amygdalus*
	−	Outer surface sculptured.	... 4
4	+	Sculpturing minute in the form of ripple marks; endocarp ovoid to oval, compressed on both sides, 25–30 mm in size, having majority of brachysclereids with some macrosclereids.	*P. armeniaca*

Contd.

Partially covered with whitish hard tissue raised above the level of seed surface	Type I	59 ± 6	Rough
In level with seed surface whitish tissue absent	Type I	89 ± 8	Rough
Almost in level with seed surface whitish tissue absent	Type III	171 ± 12	Slightly Rough
Completely covered with whitish hard tissue raised above the level of seed surface	Type I	76 ± 10	Rough
Completely covered with whitish hard tissue.	Type I	64 ± 11	Rough

Sculpturing in the form of deep ridges; endocarp ovoid, compressed on both sides, 25–45 mm in size, having majority of macrosclereids with some osteosclereids and brachysclereids. *P. persica*

3.5. WEED SEEDS

The term 'seed' in this work has been understood in its broad, popular sense. It is applied not only to true seeds, but also to equivalent structures which look like and function as seeds.

'Weed' is also a relative term, its application being partly dependent on the objectives of the person who uses it. Here the weeds have been referred to as 'plants whose virtues have not been discovered so far' and the term is interpreted in a broad sense so as to include any plant found growing wild.

The variability in the structure of seeds found throughout the angiosperms and its constancy in narrower groups permit the use of seed characteristics in the identification and classification of taxa (Martin and Barkley 1961, Musil 1963, Corner 1976). Our immediate aim in presenting the broad morphological characteristics of weed seeds here is to help archaeobotanists in particular and agriculturists, foresters, wildlife biologists, and others interested in land use in general to identify the seeds in their particular fields of interest. The descriptions and illustrations should

Table 3.8. Morpho-anatomical features of endocarps.

Species	Morphology					Anatomy			
	Shape	Size (mm)	Outer surface	Inner surface	Type of sclereids present	Frequency	Length of sclereids (μm)	Breadth of sclereids (μm)	Diameter of sclereids (μm)
Prunus persica	Oval, compressed	25–45	Deeply ridged	Smooth	Macrosclereids Osteosclereids Brachysclereids	+++ ++ ++	125–187	8.65	50–75
P. amygdalus	Ovoid to oval	30–40	Pitted	Smooth	Macrosclereids Brachysclereids	+++ ++	100–137	25–70	50–62
P. domestica	Ovoid	9–12	Smooth	Smooth	Macrosclereids Brachysclereids	+++ ++	125–375	30–60	48–52
P. cerasus	Ovoid round	17–28	Smooth	Smooth	Macrosclereids	+++	87–162	17–62	—

P. armeniaca	Ovoid	25–30	Shallow sculptured	Smooth	Brachysclereids	+++			87–100
					Macrosclereids	++	92–140	42–62	
Juglans regia	Spherical	35–45	Reticulate markings	Raised projections	Asterosclereids	+++	—	—	65–150
					Brachysclereids	+			
Zizyphus jujuba	Ovate to elliptical	8–12	With minute projections	Smooth	Macrosclereids	+++	82.5–182.5	27.5–92.5	
					Brachysclereids	+			57.5–105

+++ Very frequent; ++ Frequent; + Less frequent

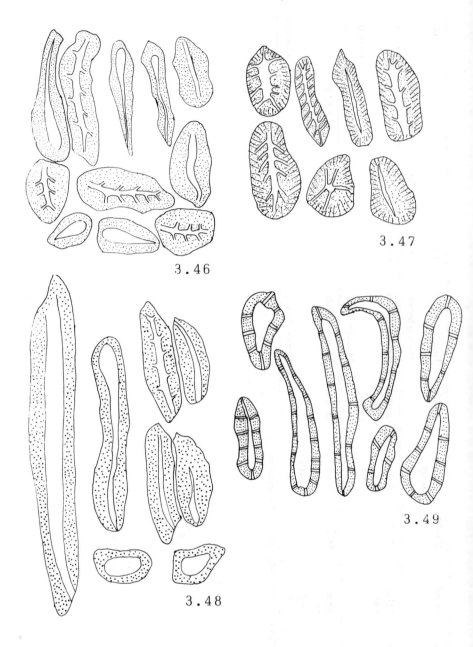

Figure 3.46: Sclereids of *Prunus persica*.
Figure 3.47: Sclereids of *P. amygdalus*.
Figure 3.48: Sclereids of *P. domestica*.
Figure 3.49: Sclereids of *P. cerasus*.

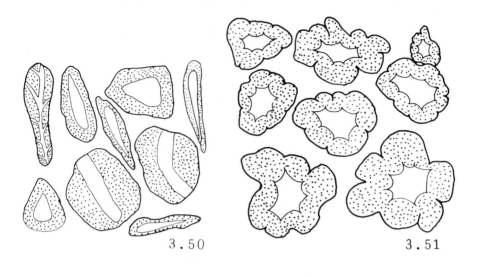
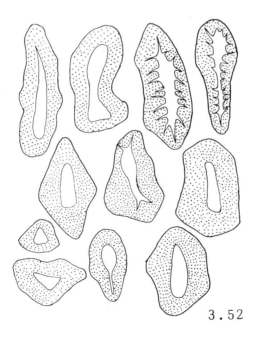

Figure 3.50: Sclereids of *P. armeniaca*.
Figure 3.51: Sclereids of *Juglans regia*.
Figure 3.52: Sclereids of *Zizyphus jujube*.

be of advantage for future research in the sense that they facilitate the identification of similar plant remains from other sites in Kashmir and adjoining regions.

The plan of description adopted here has been to arrange the seeds by family after Bentham and Hooker (1862–1883) and arrange genera and species within a family alphabetically.

RANUNCULACEAE

Adonis aestivalis Linn (fig 3.53)
Seeds triangular to oval; greenish-yellow, horny, warted, and deeply pitted; 3 mm to 6 mm in size.

Anemone biflora Dc (fig. 3.53)
Seeds oval, ovate, and flat; greenish-brown; wrinkled with small whitish hairs, 2.2 mm to 2.7 mm long and 1 mm to 1.5 mm broad.

Ceratocephalus falcatus Linn (fig. 3.53)
Seeds ovate with a long beak; greenish-yellow; smooth and sometimes seriate; 4.9 mm to 6.7 mm long and 1.7 mm to 3 mm broad.

Clematis gouriana Hook (fig. 3.53)
Seeds ellipsoid with long curved beak; crimson; smooth with dense white hairs; 5.7 mm to 7 mm long and 2 mm to 4 mm broad.

Ranunculus arvensis Linn (fig. 3.53)
Seeds ovate to elliptic, flat, beaked; brown; spiny, spines short and sharply pointed; 3 mm to 5.3 mm long and 2 mm to 2.5 mm broad.

R. laetus Wall (fig. 3.53)
Seeds elliptic with a short beak; greenish-yellow; outer surface shallow-pitted, inner surface flat; 3.2 mm to 4.1 mm long and 1.9 mm to 3 mm broad.

R. secleratus Linn (fig. 3.53)
Seeds elliptic with short beak and inner surface convex; yellowish-brown; slightly wrinkled; 1 mm to 2.3 mm long and 1 mm to 1.2 mm broad.

PAPAVERACEAE

Fumaria indica (Hausskn) Pugsley (fig. 3.53)
Seeds subglobose to globose; pale green; striated with two pits at the top; 2.3 mm to 3 mm in diameter.

Papaver macrostomum Boiss et Huet (fig. 3.53)
Seeds round to reniform; black and dark brown; surface with coarse reticulations; 0.5 mm to 1 mm long and 0.3 mm to 0.5 mm broad.

BRASSICACEAE

Arabidopsis pumila (Steph) N. Bush (fig. 3.53)
Seeds elliptic oblong; brownish-yellow; smooth, occasionally reticulate; 1 mm to 1.5 mm long and 1.2 mm broad.

Capsella bursa-postoris Moench (fig. 3.53)
Seeds elliptic ovate; brown; minutely reticulate; 1.5 mm to 1.7 mm long and 0.5 mm to 0.9 mm broad.

Coronopus didymus (Linn) Sm (fig. 3.53)
Seeds reniform to globose; black to brown; minutely reticulate; 1.5 mm to 1.7 mm long and 0.5 mm to 0.9 mm broad.

Descurainea sophia (Linn.) Webb. (fig. 3.53)
Seeds oblong oval to obovate and compressed; crimson; surface warty; 1 mm to 1.7 mm long and 0.5 mm to 0.8 mm broad.

Euclidium syriacum Br. (fig. 3.53)
Seeds spheroid to ellipsoid, greenish-yellow; shiny with minute ripples; 1 mm to 1.7 mm long and 0.7 mm to 0.9 mm broad.

Lepidium apetalum Linn (fig. 3.54)
Seeds oblong, oval to obovate; dark brown; surface granulate; 1.7 mm to 2 mm long and 1 mm broad.

L. capitatum H.ft (fig. 3.54)
Seeds oblong, oval to obovate; brown, granulose and reticulate; about 1 mm long and 0.7 mm broad.

Nasturtium officinale Br (fig. 3.54)
Seeds elliptic, oval, and winged; brownish-black; reticulate; 1 mm long and 1.3 mm broad.

Rorripa islandica (Oed) Barbas (fig. 3.54)
Seeds elliptic ovate; black to brown; reticulate; 2.7 mm to 3 mm long and 1.7 mm to 2 mm broad.

CARYOPHYLLACEAE

Cerastium glomeratum Thurill (fig. 3.54)
Seeds semi-circular to half-moon-shaped, compressed with dentate margin; brownish-yellow; reticulate; 0.6 mm to 0.9 mm long and 0.2 mm to 0.5 mm broad.

Dianthus jacquemontii Edgew (fig. 3.54)
Seeds oblong-ovate, flattish and compressed, black to brown; reticulate and dentate; 2.8 mm to 3 mm long and 1.7 mm to 2.1 mm broad.

Palaeoethnobotany

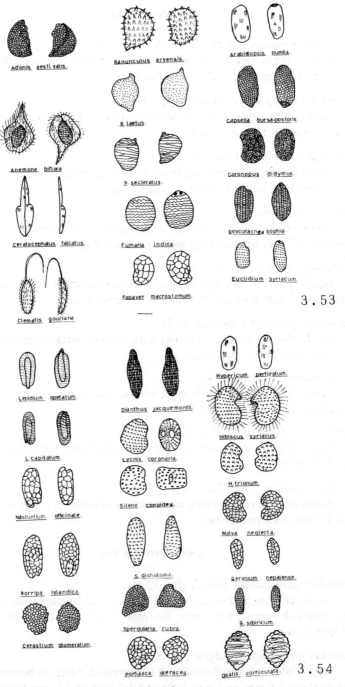

Figure 3.53–3.54: Exomorphology of weed seeds.

Lycnis coronaria Lamk. (fig. 3.54)
Seeds rounded to reniform; black; provided with tubercles in concentric rings; 0.7 mm to 1 mm in diameter.

Silene conoidea Linn (fig. 3.54)
Seeds subglobose to globose; black to brown; surface ocellate; 2.3 mm to 2.7 mm in diameter.

S. dichotoma (fig. 3.54)
Seeds oblong with pointed base and blunt apex; brown; surface rough; 1 mm to 1.7 mm long and 0.5 mm to 1 mm broad.

Spergularia rubra (Linn) J & C (fig. 3.54)
Seeds obovate, triangular, and compressed; brown; surface reticulate.

PORTULACACEAE

Portulaca oleracea Linn (fig. 3.54)
Seeds suborbicular to oval, flat and compressed; black; surface reticulate, narrowed into obscure wing; 1 mm to 1.3 mm in diameter.

HYPERICACEAE

Hypericum perforatum Linn (fig. 3.54)
Seeds minute, slender; black; surface smooth; 0.3 to 0.5 mm in size.

MALVACEAE

Hibiscus syriacus Linn (fig. 3.54)
Seeds compressed, reniform, having margin coated with brown hairs; dark brown; surface with minute ripples; 2 mm to 5 mm long and 2 mm to 3.5 mm broad.

H. trionum Linn (fig. 3.54)
Seeds suborbicular and swollen toward base; black to brown; surface papillate; 2.3 mm to 3 mm.

Malva neglecta Wall (fig. 3.54)
Seeds suborbicular to orbicular, compressed and laterally ridged; black to brown; reticulate; 2 mm.

GERANIACEAE

Geranium nepalense Sweet (fig. 3.54)
Seeds oblong to ovate; black to brown; reticulate; 2.1 mm to 2.4 mm long and 1 mm to 1.6 mm broad.

G. sibiricum Linn (fig. 3.54)
Seeds ellipsoid with a narrow line from middle of one end extending over edge with one side ridged; brown; reticulate; 1.5 mm to 1.7 mm long and 0.7 mm to 1 mm broad.

Oxalis corniculata Linn (fig. 3.54)
Seeds elliptic, flat, and transversely ridged; brown; 2 mm long and 1 mm to 1.3 mm broad.

PAPILIONACEAE

Aeschynomene indica Burn (fig. 3.55)
Seeds reniform with narrow edges; brown; surface smooth or obscurely reticulate; 2.7 mm to 3 mm long and 1.8 mm to 2 mm broad.

Astragalus grahamianus Royle (fig. 3.55)
Seeds reniform, compressed flat with rounded edges; brown; smooth; 2.8 mm to 3.2 mm long and 0.8 mm to 1.4 mm broad.

Lespedeza cuneata Don (fig. 3.55)
Seeds ovate to elliptic; greenish-yellow; surface smooth and shiny; 1.5 mm to 2 mm long and 0.5 mm to 1 mm broad.

L. juncea Pers (fig. 3.55)
Seeds ovate to elliptic; brown; surface smooth; 2.8 mm to 3 mm long and 1.3 mm to 1.7 mm broad.

L. tomentosa Sieb (fig. 3.55)
Seeds ovate to elliptic; greenish-brown; surface smooth and shiny; 2 mm to 2.4 mm long and 1 mm to 1.3 mm broad.

Medicago lupulina Linn (fig. 3.55)
Seeds ovoid, compressed and flattish, greenish-yellow to greenish-brown; surface with netted lines arising from hilum; 2 mm to 2.5 mm long and 1.5 mm to 1.7 mm broad.

M. minima (fig. 3.55)
Seeds reniform, compressed; brown; surface with longitudinal striations; 2 mm to 3 mm long and 0.7 mm to 1.7 mm broad.

M. polymorpha (fig. 3.55)
Seeds ovoid to reniform; brown; surface with longitudinal striations; 2.2 to 3.2 mm long and 0.8 mm to 1.5 mm broad.

M. sativa (fig. 3.55)
Seeds compressed, falcate; yellow to brown; surface smooth; 2 mm to 3 mm long and 1 mm to 1.5 mm broad.

Melilotus albus Lamk (fig. 3.55)
Seeds elliptic; inner surface invaginated near top; brown; surface smooth; 2 mm to 2.3 mm long and 1 mm broad.

M. indicus All (fig. 3.55)
Seeds elliptic, flat and compressed; black; surface papillate; 2 mm to 2.3 mm long and 1 mm to 1.2 mm broad.

Trifolium pratense (Linn) Podr (fig. 3.55)
Seeds compressed, ovoid, truncate at one end; yellow to brown; surface smooth; 0.8 mm to 2 mm in diameter.

T. repense (Linn) Podr (fig. 3.55)
Seeds compressed, ovoid; yellow to brown; surface smooth; 0.7 mm to 1 mm in diameter.

Vicia sativa Linn (fig. 3.55)
Seeds globose; brown with black dots; surface smooth; 1.5 mm to 5 mm in diameter.

RUTACEAE

Peganum harmala Linn (fig. 3.56)
Seeds angled triangular; dull brown; surface with polygonal markings; 2.8 mm to 3.2 mm long and 1 mm to 1.5 mm broad.

ROSACEAE

Geum urbanum Linn (fig. 3.56)
Seeds asymmetrical to reniform and flattish; reddish-brown; surface warty; 3.7 mm to 4 mm long and 2 mm to 2.7 mm broad.

Potentilla supina Linn (fig. 3.56)
Seeds elliptic ovate; brown; surface striated; 1 mm to 1.5 mm long and 1 mm broad.

Rosa macrophylla Lindl (fig. 3.56)
Seeds rounded triangular; reddish-brown surface smooth and hairy; 4.2 mm to 5.6 mm long and 2.8 to 4.2 mm broad.

R. webbiana Wall (fig. 3.56)
Seeds triangular to ovate with rounded edges; pale yellow; surface obscurely reticulate and hairy; 5.3 mm to 6.2 mm long and 3 mm to 4.5 mm broad.

Palaeoethnobotany

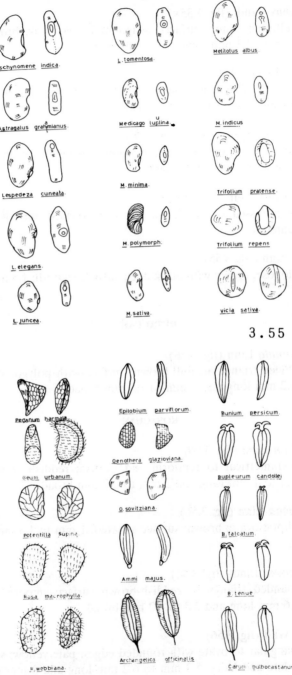

Figure 3.55–3.56: Exomorphology of weed seeds

ONAGRACEAE

Epilobium parviflorum Shreb (fig. 3.56)

Seeds oblong with outer surface smooth and inner deeply invaginated; brown; surface shallow-pitted; 0.7 mm to 1 mm long and 0.5 mm to 0.7 mm broad.

Oenothera glazioviana Linn (fig. 3.56)

Seeds triangular to elliptic ovate or oblong ovate; black to dark brown; surface reticulate to longitudinally ridged; 2 mm to 2.2 mm long and 1 mm to 1.3 mm broad.

O. sovitziana (fig. 3.56)

Seeds angled triangular; reddish-brown; surface smooth; 1.8 mm to 2.3 mm long and 1 mm to 1.5 mm broad.

APIACEAE

Ammi majus Linn (fig. 3.56)

Seeds triangular, ellipsoid to oblong; brown; surface with longitudinal ridges having a notch at the base; 1.5 mm to 2 mm long and 0.5 mm to 0.8 mm broad.

Archangelica officinalis Hoffm (fig. 3.56)

Seeds oblong and winged; yellow; surface ribbed; 5 mm to 8 mm long and 2 mm to 4 mm broad.

Bunium persicum Boiss (fig. 3.56)

Seeds ellipsoid; black; surface with longitudinal ridges; 2 mm to 2.8 mm long and 1 mm to 1.3 mm broad.

Bupleurum candollei Wall Cat (fig. 3.56)

Seeds ellipsoid to oblong, ridges prominent, furrows 3-vittate; blackish-brown; surface with longitudinal ridges; 3.5 mm to 4 mm long and 1.2 mm to 1.5 mm broad.

B. falcatum Linn (fig. 3.56)

Seeds narrowly oblong, ridges not prominent, furrows 3-vittate; brown; 1.5 mm to 2 mm long and 0.4 mm to 0.8 mm broad.

B. tenue Don. Prodr (fig. 3.56)

Seeds oblong, ridges prominent, 1-vittate; brown; surface with longitudinal ridges; 1.9 mm to 2.2 mm long and 0.5 mm to 0.9 mm broad.

Carum bulbocastanum Koch (fig. 3.56)

Seeds elliptic, convex and flattish; brown; surface with 2 to 3 ribs; 3.8 mm to 4.5 mm long and 1.1 mm to 1.8 mm broad.

Carvum carvi Linn (fig. 3.57)

Seeds oblong ovate to oblanceolate; black; smooth or with concentric hairy lines; 4 mm to 6 mm long and 2 mm to 2.5 mm broad.

Caucaulis latifolia Linn (fig. 3.57)

Seeds elliptic, plano-convex and compressed; yellow; surface having spines all over; 5 mm to 7 mm long and 2.5 mm to 3 mm broad.

Chaerophyllum villosum Wall (fig. 3.57)

Seeds oblong attenuate with brownish pappus bristles; brown; surface with longitudinal striations; 5.7 mm to 7 mm long and 1.5 mm to 2 mm broad.

Conium maculatum Linn (fig. 3.57)

Seeds oblong compressed, truncate at the top and narrow toward base; reddish-brown; surface marked with longitudinal lines; 2.3 mm to 3 mm long and 0.7 mm to 1 mm broad.

Daucus carota Linn (fig. 3.57)

Seeds elliptic, plano-convex; brown; surface with longitudinal ridges having spines; 3 mm to 4 mm long and 1.5 mm to 1.8 mm broad.

Eryngium billardieri Delar (fig. 3.57)

Seeds spheroid to ellipsoid; blackish-brown; surface smooth; 1.3 mm to 1.7 mm long and 1 mm to 1.2 mm broad.

Foeniculum vulgare Gaertn (fig. 3.57)

Seeds oval, lanceolate or elliptic; yellowish-green; surface smooth with longitudinal lines; 5 mm to 7.3 mm long and 2 mm to 2.5 mm broad.

Heracleum candicans Wall (fig. 3.57)

Seeds obovate and winged, 2-vittate; greenish-yellow; surface with longitudinal ridges; 4.3 mm to 5 mm long and 2.8 mm to 3.1 mm broad.

Hydrocotyle javanica Thunb (fig. 3.57)

Seeds reniform with outer edge rounded and inner flat; brown; surface rough; 1 mm to 1.5 mm long and 0.7 mm to 1 mm broad.

Pimpinella bella Clark (fig. 3.57)

Seeds narrowly oblong; dark brown; surface with longitudinal ridges; 1 mm to 2 mm long and 0.5 mm to 1 mm broad.

Scandix pecten-veneris Linn (fig. 3.57)

Seeds obtuse lanceolate and beaked, beak twice as long as seed; brown; surface ridged, ridges broad and spiny; 10 mm to 12 mm long and 2 mm to 4 mm broad.

Selinum tenuifolium Wall (fig. 3.57)

Seeds slender, plano-convex and winged; brown with light yellow wings; surface with longitudinal ridges; 2 mm to 4.5 mm long and 1 mm to 1.5 mm broad.

S. wallichianum (fig. 3.58)
Seeds oblong, plano-convex and winged; brown with light yellow wings; surface with longitudinal ridges; 2.8 mm to 3.3 mm long and 1.2 mm to 1.6 mm broad.

Seseli sibiricum Benth (fig. 3.58)
Seeds oblong elliptic; grey with yellowish pappus; surface with longitudinal ridges clothed with whitish hairs; 2 mm to 4 mm long and 1 mm to 1.7 mm broad.

Sium latijugum Clark (fig. 3.58)
Seeds oblong, obovate or oblanceolate compressed; blackish-brown; surface reticulate; 2.7 mm to 3.5 mm long and 1 mm to 1.5 mm broad.

Torilis japonica Prodr (fig. 3.58)
Seeds oblong, plano-convex; greyish-brown; surface spiny with long spines; 3 mm to 4.5 mm long and 1.5 mm to 1.9 mm broad.

T. leptophylla Linn (fig. 3.58)
Seeds ovate to half terrete, outer surface convex and inner deeply grooved; greenish-yellow; surface papillate; 4 mm to 4.5 mm long and 1 mm to 1.25 mm broad.

T. nodosa Linn (fig. 3.58)
Seeds oblong; greenish-yellow; surface spiny; 2.6 mm to 3.2 mm long and 1.3 mm to 1.9 mm broad.

Turgenia latifolia Hoffm (fig. 3.58)
Seeds oblong; greenish-yellow; surface with longitudinal ridges coated with spines; 5 mm to 9 mm long and 1 mm to 4 mm broad.

Vicatia conifolia Prodr (fig. 3.58)
Seeds oblong; subterrete; black to brown; outer surface smooth and inner grooved; 1.8 mm to 2.4 mm long and 0.5 mm to 0.8 mm broad.

V. wolffiana (fig. 3.58)
Seeds oblong; brown; surface with longitudinal ridges, 2.5 mm to 3 mm long and 0.9 mm to 1.2 mm broad.

RUBIACEAE

Galium aparine Linn (fig. 3.58)
Seeds globose, hollow centered on one side; greenish-brown; surface clothed with hooked bristles; 2 mm to 3.3 mm in diameter.

G. asperuloides Edgew (fig. 3.58)
Seeds globose; hollow centered on one side; dark brown; surface densely tuberculate; 2.8 mm to 3.6 mm in diameter.

Palaeoethnobotany

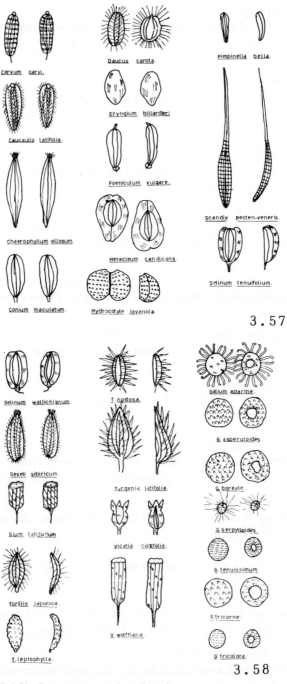

Figure 3.57–3.58: Exomorphology of weed seeds.

G. boreale Linn (fig. 3.58)
Seeds subglobose, hollow centered on one side; blackish-brown; surface coated with hairs; 0.9 mm to 1.2 mm in diameter.

G. serpylloides Royle (fig. 3.58)
Seeds minute, subglobose, hollow centered on one side; black; surface clothed with long woolly hairs; 0.3 mm to 0.5 mm in diameter.

G. tenuissimum Bieb (fig. 3.58)
Seeds minute, globose to subglobose, hollow centered on one side; dark brown; surface warted; 1 mm to 1.4 mm in diameter.

G. tricorne With (fig. 3.58)
Seeds globose, hollow centered on one side, dark brown; surface warty; 2.5 mm to 3 mm in diameter.

G. tricolora (fig. 3.58)
Seeds minute, subglobose, hollow centered on one side; reddish-brown; surface warty; 1 mm to 1.3 mm in diameter.

ASTERACEAE

Arctium lappa Linn (fig. 3.59)
Seeds oblong, cylindric with delicate brownish pappus; brown; surface with longitudinal lines; 3.5 mm to 4 mm long and 0.8 mm to 1 mm broad.

Artemesia scoparia Waldst & Kit (fig. 3.59)
Seeds spheroid to ellipsoid with whitish scar at the base; dark brown; surface with longitudinal lines; 0.9 mm to 1.1 mm in diameter.

A. tournefortiana Richb (fig. 3.59)
Seeds oblong to ovate; light brown; surface with longitudinal lines; 1 mm to 1.3 mm long and 0.6 mm to 0.7 mm broad.

Bidens biternata (Lour) Merr & Sherff (fig. 3.59)
Seeds linear and several-sided to flat oblanceolate, four barbed awns extend upward from top; black; surface with longitudinal lines; 12 mm to 16 mm long and 1 mm broad.

B. cernua Linn (fig. 3.59)
Seeds 4-angled, broad at the top, which is truncate and hairy; white to greenish; rough; 3 mm to 4.3 mm long and 1.7 mm to 2.3 mm broad.

Carduus edelbergii Rech (fig. 3.59)
Seeds linear; golden shiny; warted; 3.5 mm to 5 mm long and 2.5 mm to 3 mm broad.

C. nutans Linn (fig. 3.59)
Seeds elongate, oblong oval, compressed, top circular with flask-shaped cap; yellow; surface warted; 4.5 mm to 5 mm long and 2 mm to 2.5 mm broad.

Carpesium abrotanoides Linn (fig. 3.59)
Seeds linear oblong; pale brown; surface striated; 2.7 mm to 3 mm long and 0.5 mm to 0.7 mm broad.

Causinia microcarpa Boiss (fig. 3.59)
Seeds oblong, ovate, compressed and flattened on inner side and convex on outer side; brick red with black dots; smooth; 2.1 mm to 2.4 mm long and 1.3 mm to 1.7 mm broad.

Cichorium intybus Linn (fig. 3.59)
Seeds compressed and arched oblong; top truncate, body narrow toward base; brown; surface scaly; 1.5 mm to 2 mm long and 1 mm broad.

Circium wallichii Dc (fig. 3.59)
Seeds oblong, ovate, elongated, compressed, and truncate at the top; brownish-black; surface sculptured; 2 mm to 2.3 mm long and 1.3 mm to 1.6 mm broad.

Crepis sancta ssp. *bifida* Babc (fig. 3.59)
Seeds long, flat, angular, or ridged; brown, yellow toward the ends; surface with cross and longitudinal lines; 3.4 mm to 5.3 mm long and 0.4 mm to 1 mm broad.

Erigeron alpinus Linn (fig. 3.59)
Seeds minute, slender with a pappus of white long hairs; greenish-yellow to brown; surface hairy; 2 mm to 3 mm long and 0.5 mm to 0.7 mm broad.

E. bonariensis Linn (fig. 3.60)
Seeds minute, slender with a pappus of white hairs; yellowish-brown; surface with dense hairs; 0.5 mm to 0.9 mm long and 0.2 mm to 0.3 mm broad.

E. canadensis Linn (fig. 3.60)
Seeds linear oblong; dark brown; surface smooth; 1 mm to 1.2 mm long and 0.5 mm to 0.7 mm broad.

E. multicaulis Wall (fig. 3.60)
Seeds linear, oblong and angled with a long white pappus; light brown; surface hairy; 1 mm to 1.2 mm long and 0.5 mm to 0.6 mm broad.

Hieracium umbellatum Linn (fig. 3.60)
Seeds cylindrical with truncate top; reddish; surface ribbed; 3 mm to 3.5 mm long and 0.5 mm to 1 mm broad.

Keys and Criteria for Identification

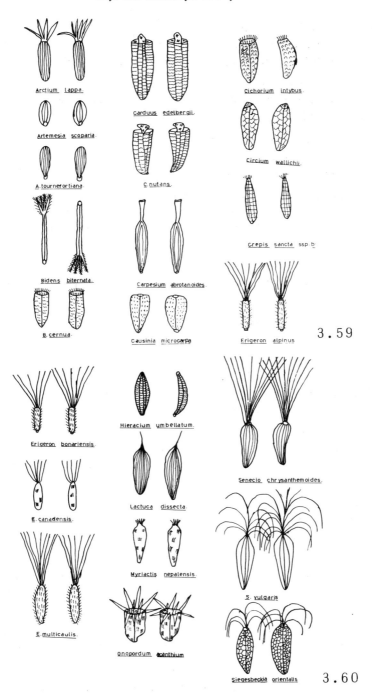

Figure 3.59–3.60: Exomorphology of weed seeds

Lactuca dissecta D. Don. (fig. 3.60)

Seeds oblanceolate, beaked and compressed; light brown; surface ribbed; 5.3 mm to 6.9 mm long and 1 mm to 1.3 mm broad.

Myriactis nepalensis Less (fig. 3.60)

Seeds oblanceolate and compressed; brown; surface smooth; 2 mm to 2.5 mm long and 0.6 mm to 0.9 mm broad.

Onopordum acanthium (fig. 3.60)

Seeds oblong and four-sided, crowned with several tapering pappus scales; cream; smooth; 4 mm to 6 mm long and 2 mm to 4 mm broad.

Senecio chrysanthemoides Prodr (fig. 3.60)

Seeds slender, oblong with white pappus; brown; surface with longitudinal ridges; 1.5 mm to 4 mm long and 0.3 mm to 1 mm broad.

S. vulgaris Linn (fig. 3.60)

Seeds ovate, oblong, flat and compressed with white woolly pappus; light brown; surface with 2 to 3 ribs; 2.8 mm to 3.2 mm long and 1 mm to 1.4 mm broad.

Siegesbeckia orientalis Linn (fig. 3.60)

Seeds obovate, cuneate, 4-angled; shiny black; surface reticulate; 3 mm to 3.2 mm long and 1 mm to 1.3 mm broad.

Sonchus arvensis Linn (fig. 3.61)

Seeds ellipsoid and compressed; brown to black; surface transversely rugose; 1.8 mm to 3.6 mm long and 1 mm to 1.3 mm broad.

S. asper (fig. 3.61)

Seeds ovate, oblong, compressed, brown; with long white woolly pappus; surface with longitudinal ridges; 2 mm to 2.8 mm long and 1 mm to 1.3 mm broad.

S. oleraceus Linn Boiss (fig. 3.61)

Seeds compressed ellipsoid, 3-ribbed; brick red; 2 mm to 3.5 mm long and 0.8 mm to 1.5 mm broad.

Taraxacum officinale Wigg (fig. 3.61)

Seeds flattish, oblanceolate or 4-angled, narrow at both ends; greyish-yellow; surface ribbed; 3.8 mm to 4.2 mm long and 1.6 mm to 2.2 mm broad.

Tragopogon kashmirianus G. Singh (fig. 3.61)

Seeds slender, lanceolate and many-ribbed with a pappus of numerous hairs; buff; surface ribbed; 1.7 mm to 2.3 mm long and 2 mm to 3 mm broad.

Valerianella muricata Steu Boixt (fig. 3.61)

Seeds ovoid with a wing at the base and swollen apex; dark brown; surface hairy; 2 mm long and 1.5 mm broad.

BORAGINACEAE

Cynoglossum glochidiatum Wall (fig. 3.62)
Seeds ovate to globose, plano-convex; black; surface toothed; 2.5 mm to 3 mm long and 1.2 mm to 1.5 mm broad.

Lithospermum arvense Linn (fig. 3.62)
Seeds ovoid truncate with a ridge on one side and broad flattish scar at base; brownish-black; surface coarsely rough and spiny; 2 mm to 4 mm long and 1 mm to 1.7 mm broad.

CONVOLVULACEAE

Convolvulus arvensis Linn (fig. 3.62)
Seeds rounded triangular and obtuse; black; surface obscurely papillate; 3 mm to 4 mm long and 1.8 mm to 2 mm broad.

Cuscuta chinesis Lamk (fig. 3.62)
Seeds globose to rounded, triangular; blackish-brown; warty surface.

Ipomoea eriocarpa Br Poda (fig. 3.62)
Seeds rounded triangular; black to brown; warty surface; 2.4 mm to 3 mm long and 1.7 mm to 2 mm broad.

I. hispida R & S (fig. 3.62)
Seeds triangular ovate; light brown; surface obscurely smooth; 3 mm to 3.2 mm long and 2 mm to 2.2 mm broad.

I. palmata Forsk (fig. 3.62)
Seeds globose; ovoid rounded; inner surface flat; greyish-black; surface smooth; 6.8 mm to 8 mm long and 3.9 mm to 6.1 mm broad.

SOLANACEAE

Datura stramonium Linn (fig. 3.62)
Seeds compressed oval to reniform, plano-convex; black; surface with reticulate markings; 2.8 mm to 3.7 mm long and 2 mm to 2.5 mm broad.

Solanum miniatum Bernh (fig. 3.62)
Seeds ovate, obtuse, compressed and flattened; brown; surface reticulate; 1.5 mm to 2.3 mm long and 0.5 mm to 1 mm broad.

SCROPHULARIACEAE

Linaria palmatica Linn (fig. 3.62)
Seeds rounded, reniform; black; surface warty; 0.5 mm to 1.5 mm in diameter.

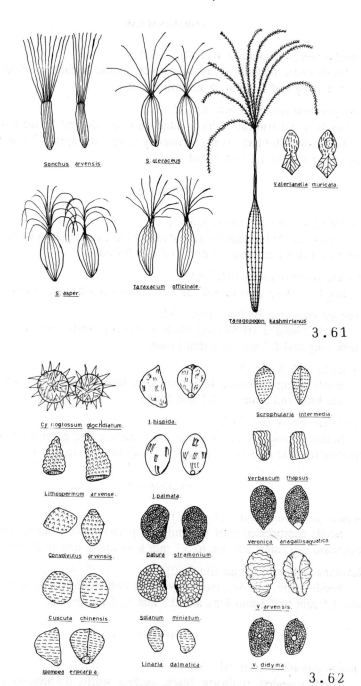

Figure 3.61–3.62: Exomorphology of weed seeds

Scrophularia intermedia (fig. 3.62)
Seeds ellipsoid; black; surface rough; 1 mm to 1.5 mm long and 0.3 mm to 0.7 mm broad.

Verbascum thapsus Linn (fig. 3.62)
Seeds cylindric, truncate with wavy longitudinal ridges; blackish-brown; surface with longitudinal ridges; 1 mm to 1.5 mm long and 0.3 mm to 0.6 mm broad.

Veronica anagallisaquatica Linn (fig. 3.62)
Seeds oval, oblong or oblong ovate and compressed; yellow to brown; surface glossy and reticulate; 0.4 mm to 0.8 mm in diameter.

V. arvensis Linn (fig. 3.62)
Seeds elliptic, ovate or oval, outer surface convex and inner invaginated; light yellow; surface translucent and transversely ridged; 1.6 mm to 1.9 mm long and 0.4 mm to 0.9 mm broad.

V. didyma Linn (fig. 3.62)
Seeds obovate, oblong or ovate, compressed with outer surface convex and inner concave; golden yellow; surface translucent and reticulate; 1 mm to 1.1 mm long and 0.5 mm to 0.7 mm broad.

V. persica Poir (fig. 3.63)
Seeds elliptic oval to ovate; outer surface convex and inner deeply invaginated; brown; surface granular; 1.7 mm to 2.0 mm long and 0.7 mm to 0.9 mm broad.

VERBENACEAE

Verbena officinalis Linn (fig. 3.63)
Seeds oblong; dark brown; outer surface ridged and inner fairly reticulate; 1.5 mm to 2.1 mm long and 0.5 mm to 1.0 mm broad.

LAMIACEAE (LABIATAE)

Clinopodium umbrosum Hook (fig. 3.63)
Seeds compressed, globose; dark brown; smooth; 0.5 mm to 0.8 mm in diameter.

C. vulgare Linn (fig. 3.63)
Seeds orbicular, sometimes globose, often compressed; brown; smooth; 1 mm to 1.5 mm in diameter.

Elsholtzia cristata Willd (fig. 3.63)
Seeds obovate-oblong, compressed and flattened; brown; surface reticulate; 1 mm to 1.4 mm in size.

E. densa Benth (fig. 3.63)

Seeds obovate-oblong, obovate; compressed; brown; surface obscurely granulate; 1.3 mm to 1.7 mm in size.

Lamium amplexicaule Linn (fig. 3.63)

Seeds oblong-obovate, 3-angled, plano-convex; brown; surface mottled; 2.9 mm to 3.5 mm long and 1 mm to 1.3 mm broad.

Lycopus europaeus Linn (fig. 3.63)

Seeds obovate-oblong, compressed and flat; brown; surface smooth and grooved; 2 mm to 2.2 mm long and 1 mm broad.

Marrubium vulgare Linn (fig. 3.63)

Seeds ellipsoid triangular; greyish with black dots; smooth; 1.7 mm to 2.5 mm long and 1 mm to 1.7 mm broad.

Nepeta cataria Linn (fig. 3.63)

Seeds broadly oblong, triangular with rounded edges; dark brown; surface smooth; 1 mm to 1.9 mm long and 0.7 mm to 1 mm broad.

Ocimum sanctum Linn (fig. 3.63)

Seeds obovate, oblong; compressed and flat; black, shiny; surface obscurely reticulate 1.5 mm to 1.7 mm long and 0.5 mm to 0.7 mm broad.

Organum vulgare (fig. 3.63)

Seeds minute, spheroid to ovoid, black and brown; surface smooth, obscurely hairy; 1 mm to 1.5 mm in diameter.

Salvia moorcroftiana Wall (fig. 3.63)

Seeds subglobose; creamy with brown longitudinal lines; surface smooth; 1.7 mm to 2.5 mm in diameter.

Stachys floccosa Benth (fig. 3.63)

Seeds ovate, elliptic to oblong with pointed apex; black; surface with minute reticulations; 0.5 mm to 1.5 mm long and 1.8 mm to 2.3 mm broad.

PLANTAGINACEAE

Plantago lanceolata Linn (fig. 3.63)

Seeds plano-convex and elliptic; brown; smooth; 1 mm to 2 mm long and 0.5 mm to 0.5 mm broad.

P. major Linn (fig. 3.64)

Seeds elliptic-triangular, edges rounded or narrowed into wings and flat; dark brown; smooth; 0.7 mm to 1.7 mm long and 0.3 mm to 0.7 mm broad.

AMARANTHACEAE

Amaranthus blitum Linn (fig. 3.64)
Seeds compressed, globose to subglobose; shiny black; smooth; 0.7 mm to 1 mm in diameter.

A. caudatus Linn (fig. 3.64)
Seeds circular, lenticular, flattened and compressed; shiny black to reddish-brown; surface with minute reticulations; 1 mm to 1.3 mm in diameter.

A. hybridus Linn (fig. 3.64)
Seeds small, rounded; yellowish; smooth; 0.7 mm to 0.9 mm in diameter.

A. virdis Linn (fig. 3.64)
Seeds rounded, flattish and circular; dark brown; smooth; 1 mm to 1.7 mm in diameter.

CHENOPODIACEAE

Chenopodium album Linn (fig. 3.64)
Seeds circular, lenticular, compressed with rounded edges; shiny black to brown; surface reticulate, 1 mm to 1.1 mm.

C. blitum Hook (fig. 3.64)
Seeds circular, lenticular; black; surface warty; 0.7 mm to 1 mm in diameter.

C. foliosum Schrad (fig. 3.64)
Seeds ovate to elliptic-ovate, flattened with rounded edges; black; surface smooth and shiny; 1.3 mm to 1.5 mm.

C. glaucum Linn (fig. 3.64)
Seeds oval to elliptic-ovate, flattened with rounded edges; black; surface obscurely granular; 1.4 mm to 1.6 mm.

POLYGONACEAE

Fagopyrum esculentum Moench (fig. 3.64)
Seeds sharp-edged, triangular, marked with zebra-like streaks; greyish-black; surface rough; 4.5 mm to 5.2 mm long and 3.5 mm to 3.7 mm broad.

Polygonum convolvulus Linn (fig. 3.64)
Seeds elliptic, oval, oblong or oblong-ovate; shining black; surface smooth; 3 mm to 3.2 mm long and 0.9 mm to 1.9 mm broad.

Palaeoethnobotany

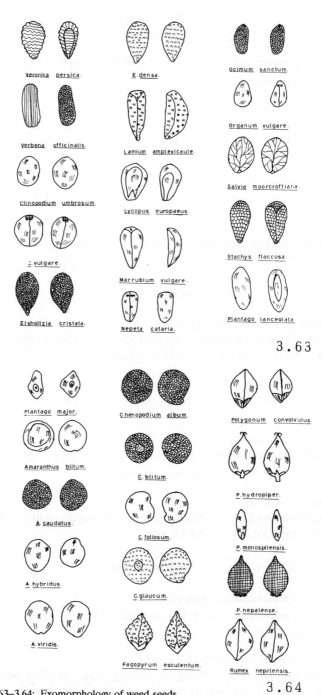

Figure 3.63–3.64: Exomorphology of weed seeds.

P. hydropiper Linn (fig. 3.64)
Seeds elliptic-ovate, flat and shortly beaked; black to brown; surface smooth; 3 mm to 3.2 mm long and 0.9 mm to 1.9 mm broad.

P. monospliensis (fig. 3.64)
Seeds elliptic; brown; surface smooth; 0.3 mm to 0.6 mm long and 0.2 mm to 0.3 mm broad.

P. nepalense Meissn (fig. 3.64)
Seeds elliptic-ovate, flat and shortly beaked; black to brick red; surface coarsely reticulate; 1.7 mm to 2.5 mm long and 0.9 mm to 1.2 mm broad.

Rumex nepalensis Spr (fig. 3.64)
Seeds sharply triangular and acute-pointed at the apex; shiny; brown, smooth; 2.7 mm to 3.5 mm long and 1.7 mm to 2.3 mm broad.

EUPHORBIACEAE

Euphorbia helioscopia Linn (fig. 3.65)
Seeds oblong with distinct longitudinal lines; yellow to creamy brown; surface shallow-pitted; 1 mm to 1.2 mm long and 0.7 mm to 0.9 mm broad.

E. kanoarica Boiss (fig. 3.65)
Seeds oblong; dirty green; smooth; 4.5 mm to 5.3 mm long and 2 mm to 2.5 mm broad.

E. peplus Linn (fig. 3.65)
Seeds oblong with distinct longitudinal lines; grey; 2.8 mm to 3.2 mm long and 0.9 mm to 1.7 mm broad.

E. prostrata Torr (fig. 3.65)
Seeds oblong; creamy white; surface reticulate with distinct furrows; 1 mm to 1.3 mm long and 0.4 mm to 0.9 mm broad.

E. tibetica Boiss (fig. 3.65)
Seeds oblong-ovate or oblong-obvate with a distinct longitudinal line; dirty green; smooth; 4.2 mm to 5.3 mm long and 2 mm to 2.2 mm broad.

IRIDACEAE

Iris ensata Thumb (fig. 3.65)
Seeds ellipsoid, 2-vittate with one convex side and other concave; body black, wings reddish-brown; surface with minute tubercles; 2.7 mm to 3 mm long and 2 mm to 2.5 mm broad.

CANNABACEAE

Cannabis sativa Linn (fig. 3.65)
Seeds globose; greyish with brownish-black striations; surface smooth; 2 mm to 3.5 mm in diameter.

LILIACEAE

Asparagus filicinus Ham. (fig. 3.65)
Seeds rounded and compressed with a whitish scar in the centre; black; reticulate; 3.7 mm to 4 mm in diameter.

A. officinalis Linn (fig. 3.65)
Seeds lanceolate; black; ridged surface; 2 mm to 3 mm long and 0.5 mm to 0.7 mm broad.

CYPERACEAE

Cyperus iria Linn (fig. 3.65)
Seeds 3-angled, lens-shaped or ovate, shortly beaked; brown; surface obscurely granulate; 0.9 mm to 1.3 mm long and 0.3 mm to 0.7 mm broad.

C. rotundus Linn (fig. 3.65)
Seeds triangular, compressed, beaked; black to brown; granulate; 0.7 mm to 1 mm long and 0.5 mm to 0.7 mm broad.

C. serotinus Rottb (fig. 3.65)
Seeds obovate to ovate-oblong, compressed with a distinct beak; brown; coarsely reticulate; 1.2 mm to 1.7 mm long and 0.4 mm to 1 mm broad.

POACEAE

Aegilops tauschii Cors (fig. 3.65)
Seeds obovate or ovate-oblong, ventrally grooved and hairy; yellowish-brown; smooth; 3.8 mm to 5.1 mm long and 1.5 mm to 1.7 mm broad.

Bromus japonicus Thumb (fig. 3.65)
Seeds obovate; upper surface brown and lower white; surface with longitudinal ridges; 5.7 mm to 6 mm long and 1.3 mm to 1.7 mm broad.

Digitaria sanguinolentus Edgew (fig. 3.65)
Seeds ovate, elliptic or obovate; pale yellow or greenish-yellow; surface smooth or obscurely granulate; 1.9 mm to 2.6 mm long and 1 mm to 1.4 mm.

Echinochloa crusgalli Beauv (fig. 3.66)
Seeds ovate, lanceolate or ovate-oblong to boat-shaped; greenish-yellow; surface smooth; 2.9 mm to 3.4 mm long and 1 mm to 1.7 mm broad.

Lolium temulentum Linn (fig. 3.66)
Seeds elliptic-ovate or boat-shaped; pale green; surface smooth or obscurely granulate; 1.7 mm to 2.5 mm long and 0.9 mm to 1.5 mm broad.

Phleum paniculatum Linn (fig. 3.66)
Seeds ovate, lanceolate, ovate-oblong or linear; pale green to yellow; smooth; 1 mm to 1.5 mm long and 0.7 mm to 1 mm broad.

Poa bulbosa Linn (fig. 3.66)
Seeds ovate-lanceolate or ovate-acute; green; smooth; 1 mm to 1.6 mm long and 0.4 mm to 0.6 mm broad.

P. pratensis Linn (fig. 3.66)
Seeds ovate-lanceolate, acute with rounded edges; pale green; smooth; 0.7 mm to 1.2 mm long and 0.4 mm to 0.6 mm broad.

Polypogon fugax Nees ex Steud (fig. 3.66)
Seeds ovate-oblong with rounded tip; pale green; smooth; 1 mm to 1.3 mm long and 0.7 mm to 0.8 mm broad.

P. monspeliensis Desf. (fig. 3.66)
Seeds obovate, ovate-oblong, ovate or elliptic; pale green to light green; smooth; 0.9 mm to 1 mm long and 0.5 mm to 0.7 mm broad.

Setaria glauca Beauv (fig. 3.66)
Seeds elliptic-ovate to oval, dorsal surface convex; brown; dorsal surface transversely ridged and inner smooth, 2.8 mm to 4 mm long and 1 mm to 1.5 mm broad.

S. viridis Beauv (fig. 3.66)
Seeds elliptic-ovate to oval; pale to light green; dorsal surface transversely ridged; 3 mm to 3.5 mm long and 1.2 mm to 1.5 mm broad.

Sorghum halepense Wall (fig. 3.66)
Seeds elliptic-lanceolate; pale green to light green; dorsal surface hairy and ventral grooved; 4.5 mm to 5 mm long and 1.5 mm to 1.9 mm broad.

From the description and sketches of various species it is evident that genera and species vary considerably in their seed morphology and can be conveniently identified with reference to the present study.

3.6 WOODS (FIG. 3.67–3.88)

Charcoal is usually the most abundant archaeological plant material. To identify the archaeological charcoals correctly, anatomy of some woods

Palaeoethnobotany

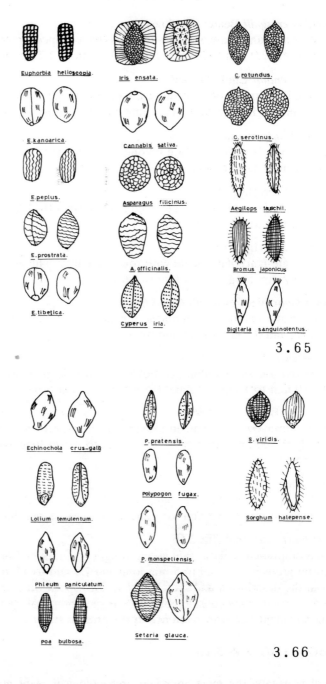

Figure 3.65–3.66: Exomorphology of weed seeds.

Keys and Criteria for Identification   73

growing in Kashmir has been worked out. Anatomical features of soft woods (gymnosperms) are summarized in table 3.9 and those of hard wood (angiosperms) in table 3.10. Based on these characters a key for the identification of various genera is as follows:

KEY TO THE IDENTIFICATION OF SOME WOODS FROM KASHMIR

1	+	Wood without vessels (non-porous).	... 2
	−	Wood with vessels (porous).	... 8
2	+	Vertical parenchyma absent or rarely present.	... 3
	−	Vertical parenchyma usually present in abundance.	... 7
3	+	Ray tracheids normally present.	... 4
	−	Ray tracheids normally absent.	... 6
4	+	Vertical and horizontal resin canals usually present.	... 5
	−	Vertical and horizontal resin canals usually absent; traumatic ducts may be present; rays tall, often more than 30 cells in height.	*Cedrus*
5	+	Resin canals with thin-walled epithelial cells; cross field pits window-like to pinoid; early wood tracheids without spiral thickenings.	*Pinus*
	−	Resin canals with thick-walled epithelial cells; cross field pits piceiod type; early wood tracheids with spiral thickenings.	*Picea*
6	+	Tracheids with prominent spiral thickenings; horizontal walls of ray cells not pitted; end walls nondulated; cross field pits cupressoid type.	*Taxus*
	−	Tracheids without spiral thickenings, horizontal walls of ray cells highly pitted, end walls nondulated, cross field pits taxodiod type.	*Abies*
7	+	Vertical parenchyma zonate, early wood tracheids with prominent intercellular spaces.	*Juniperus*
	−	Vertical parenchyma not zonate; early wood tracheids with less intercellular spaces; rays may be occasionally biseriate; cross field pits cupressoid.	*Cupressus*
8	+	Spring wood vessels larger than those in the summer wood, i.e., wood ring porous.	... 9
	−	Spring wood vessels slightly or not larger than those in summer wood, i.e., wood diffuse porous.	... 21
9	+	Broad rays of oak type present.	*Quercus*
	−	Broad rays of oak type absent.	... 10
10	+	Spring wood vessels grading in size into those of summer wood, i.e., wood semi-ring porous.	... 11
	−	Spring wood vessels not grading in size into those of summer wood, i.e., wood typically ring porous.	... 15
11	+	Rays 1-6 seriate; parenchyma included in the body of the growth ring or wanting.	... 12
	−	Rays uniseriate; parenchyma terminal.	... 14

12	+	Parenchyma sparse; vessels with spiral thickenings, largest vessels usually less than 100 μm in diameter.	*Prunus*
	−	Parenchyma relatively abundant, vessels without spiral thickening or spirals restricted to summer wood vessels, largest vessels more than 100 μm in diameter.	... 13
13	+	Summer wood vessels with spiral thickening, those in outer portion of the growth ring associated with parenchyma forming tangential, several-seriate, more or less continuous bands.	*Catalpa*
	−	Summer wood vessels without spiral thickening, solitary or in radial groups of 2-4; parenchyma metatracheal, sometimes crystalliferous, ray cells round to elliptic; fibres 25 μm in diameter.	*Juglans*
14	+	Rays essentially heterogeneous.	*Salix*
	−	Rays essentially homogeneous.	*Populus*
15	+	Summer wood figured with concentric, wavy, more or less continuous bands of vessels.	... 16
	−	Summer wood without concentric bands of vessels or occasionally with interrupted bands in the outer portion of the growth ring consisting of vessels and appreciable amounts of parenchyma.	... 18
16	+	Spring wood vessels in one row.	*Ulmus*
	−	Spring wood vessels in several rows.	... 17
17	+	Rays 1-6 seriate, essentially homogeneous.	*Ulmus*
	−	Rays 1-13 seriate, essentially heterogeneous.	*Celtis*
18	+	Rays uniseriate or rarely in part biseriate; spring wood vessels in several rows.	*Castanea*
	−	Rays uniseriate to many-seriate.	... 19
19	+	Rays 1-4 seriate.	... 20
	−	Rays 1-5+ seriate, spring wood vessels occluded with tyloses, inter-vessel pits vestured.	*Robinia*
20	+	Vessels with simple perforation plates, summer wood vessels without spiral thickening, solitary or in rows of two to several, fibres with a maximum diameter of less than 25 μm.	*Fraxinus*
	−	Spring wood vessels partly or completely occluded with tyloses, parenchyma paratracheal, aliform confluent.	*Morus*
21	+	Wood semi-diffuse porous.	... 22
	−	Wood typically diffuse porous.	... 25
22	+	Rays uniseriate to many-seriate, parenchyma present in the body of the ring or wanting.	... 23
	−	Rays uniseriate; parenchyma confined to the outer margin of the ring.	... 24
23	+	Longitudinal parenchyma extremely sparse or wanting, vessels with spiral thickening, largest less than 100 μm in diameter.	*Prunus*
	−	Longitudinal parenchyma relatively abundant, vessels	

Keys and Criteria for Identification

		without spiral thickening, largest more than 100 μm in diameter, parenchyma may be crystalliferous, ray cells round to elliptic.	*Juglans*
24	+	Rays essentially homogeneous.	*Populus*
	–	Rays essentially heterogeneous.	*Salix*
25	+	Widest rays less than 8 seriate.	... 26
	–	Widest rays more than 8 seriate. Rays 1-14 seriate, inter-vessel pits not crowded.	*Platanus*
26	+	Rays of two types; simple and aggregate.	... 27
	–	Rays of one type; simple only.	... 28
27	+	Simple rays uniseriate or rarely biseriate, inter-vessel plates scalariform, inter-vessel pits not crowded.	*Alnus*
	–	Simple rays, 1-4 seriate, perforation plates, simple, inter-vessel pits crowded.	*Carpinus*
28	+	Vessels solitary or in more or less irregular groups of two to several, generally more numerous and crowded in spring wood.	*Prunus/Pyrus*
	–	Vessels solitary or in radial rows of 2 to 5.	... 29
29	+	Largest vessels 40 to 60 μm in diameter, with or without spiral thickenings; parenchyma sometimes restricted to few cells only; rays uniseriate or biseriate.	*Aesculus*
	–	Largest vessels more than 60 μm in diameter, without spiral thickenings.	... 30
30	+	Vessels uniformly distributed; rays of two widths: narrow 1-3 seriate and broad 3-8 seriate.	... 31
	–	Rays not of two widths.	... 33
31	+	Perforation plates simple or scalariform, rays homo- to heterogeneous, pits on vessel walls orbicular to hexagonal.	*Betula*
	–	Vessels numerous; rays essentially homogeneous, parenchyma sparse.	... 32
32	+	Rays essentially uniseriate and occasionally biseriate, vessels usually rounded.	*Parrotiopsis*
	–	Rays 2-3 seriate, essentially homogeneous; vessels elliptic to oval.	*Crataegus*
33	+	Rays 3-5 seriate, essentially homogeneous, parenchyma paratracheal and apotracheal, vessels 80-120 μm in diameter.	*Acer*
	–	Rays 1-2 seriate, heterogeneous, parenchyma apotracheal, sparse, vessels 40-75 μm in diameter.	*Viburnum*

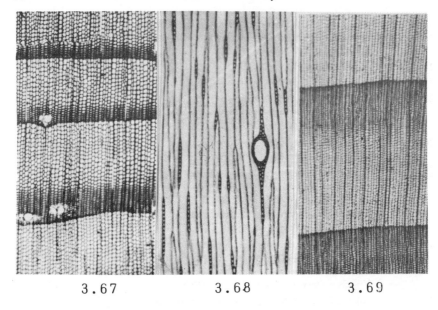

Figure 3.67: Transverse section (T.S.) *Picea smithiana* 25x.
Figure 3.68: Tangential longitudinal section (T.L.S) *Pinus wallichiana* 50x.
Figure 3.69: T.S. *Abies pindrow* 25x.

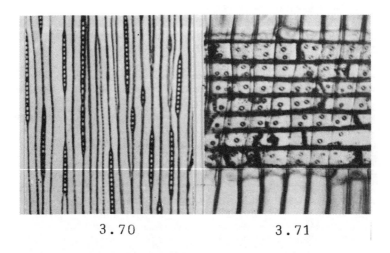

Figure 3.70: T.L.S. *Cedrus deodara* 50x.
Figure 3.71: Radial longitudinal section (R.L.S) *Cedrus deodara* 250x.

Keys and Criteria for Identification

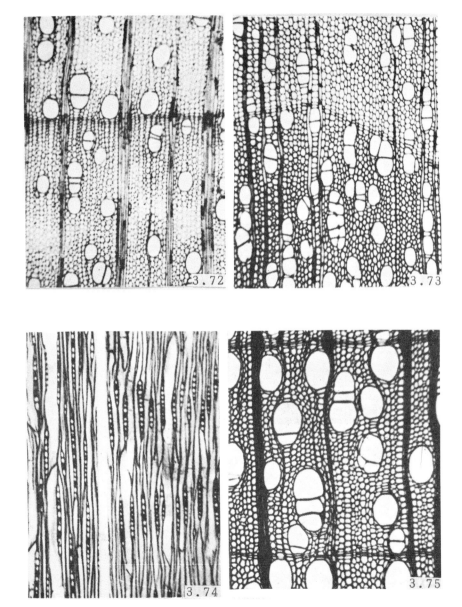

Figure 3.72: T.S. *Acer* sp. 50x.
Figure 3.73: T.S. *Aesculus indica* 50x.
Figure 3.74: T.L.S. *Aesculus indica* 50x.
Figure 3.75: T.S. *Betula utilis* 50x.

78 *Palaeoethnobotany*

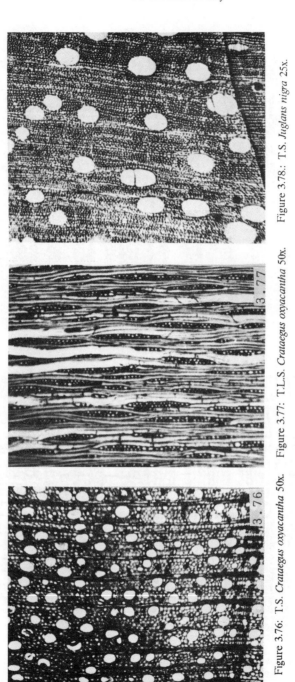

Figure 3.76: T.S. *Crataegus oxyacantha* 50x. Figure 3.77: T.L.S. *Crataegus oxyacantha* 50x. Figure 3.78: T.S. *Juglans nigra* 25x.

Keys and Criteria for Identification 79

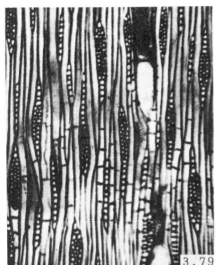
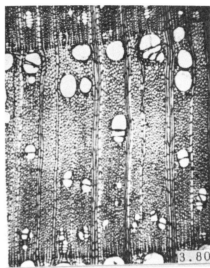
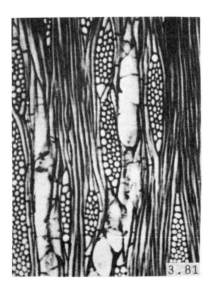
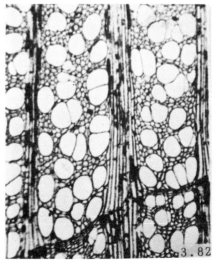

Figure 3.79: T.L.S. *Juglans nigra* 50x.
Figure 3.80: T.S. *Morus alba* 25x.
Figure 3.81: T.L.S. *Morus alba* 50x.
Figure 3.82: T.S. *Platanus orientalis* 50x.

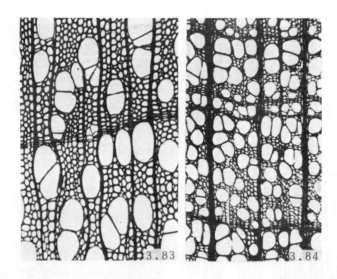
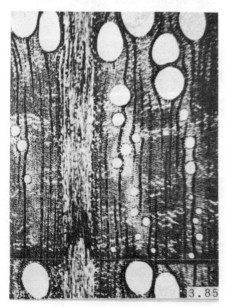
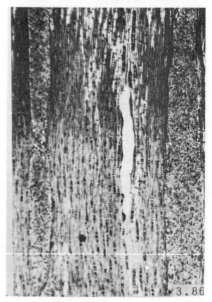

Figure 3.83: T.S. *Populus* sp. 50x.
Figure 3.84: T.S. *Prunus* sp. 50x.
Figure 3.85: T.S. *Quercus robur* 25x.
Figure 3.86: T.L.S. *Quercus robur* 25x.

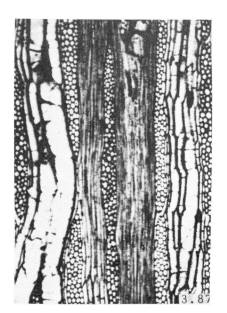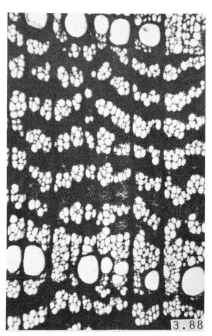

Figure 3.87: T.L.S. *Robinia pseudoacacia* 50x.
Figure 3.88: T.S. *Ulmus wallichiana* 25x.

Table 3.9. Anatomy of some soft

Species	Longitudinal tracheids		Longitudinal parenchyma	
	Average diameter (μm)	Spiral thickenings	Present/ wanting	Nature
Abies pindrow	32	Wanting	Wanting	—
Cedrus deodara	35	Wanting	Wanting	—
Cupressus spp.	24	Wanting	Present	Abundant, scattered
Juniperus recurva	23	Wanting	Present	Solitary cells or in bands of 2-3
Picea smithiana	28	Wanting	Wanting	—
Pinus wallichiana	29	Wanting	Wanting	—
Taxus baccata	19	Present	Wanting	—

woods of Kashmir

Longitudinal resin/ traumatic canals			Wood rays				Cross field pitting	
Present/ wanting	Nature of epithelial cells	Maximum tangential diameter (μm)	Seriation	Composition	Height		Number per cross Nature	field
					in cells	in μm		
Wanting	—	—	Uniseriate	Homogeneous	2-18	15-430	Taxodioid	2-3
Traumatic canals present	—	75	Uniseriate	Heterogeneous	4-32	16-536	Taxodioid or piceioid	2-4
Wanting	—	—	Uniseriate	Homogeneous	1-15	6-210	Cupressoid	2-3
Wanting	—	—	Uniseriate	Homogeneous	1-8	6-170	Cupressoid	2-4
Resin canals present	Thick-walled	135	Uniseriate and fusiform	Heterogeneous	2-20	8-345	Piceioid	2-4
Resin canals present	Thin-walled	170	Uniseriate and fusiform	Heterogeneous	1-15	6-245	Large, window-like to pinoid	1-3
Wanting	—	—	Uniseriate	Homogeneous	1-25	5-230	Cupressoid	1-4

Table 3.10 Anatomy of some

Species	Topography of wood	Vessels						Parenchyma	
		Arrangement of pores	Size (μm)	Spiral thickenings	Inter vessel pitting	Nature of perforations	Inclusions	Arrangement	Inclusions
1	2	3	4	5	6	7	8	9	10
Acer spp.	Diffuse porous	Solitary or in multiples of 2 or more	80–120	Present	Orbicular to angular	Simple	Wanting	Paratracheal and apotracheal	Wanting
Aesculus indica	Diffuse porous	Solitary or in multiples of 2 or more	30–80	Occasionally present	Orbicular to oval or angular	Simple	Wanting	Terminal sparse	Wanting
Alnus nitida	Diffuse porous	Solitary and in multiples of 2–3	70–110	Wanting	Orbicular to oval	Scalariform	Wanting	Metatracheal diffuse	Wanting
Betula utilis	Diffuse porous	Solitary or in multiples of 2 or more	40–160	Wanting	Orbicular to broad oval	Scalariform	Wanting	Metatracheal diffuse and paratracheal	Wanting
Carpinus spp.	Diffuse porous	Multiples of 2 or more	60–115	Wanting	Orbicular to angular	Scalariform	Wanting	Metatracheal diffuse	Wanting

hard woods of Kashmir

Fibres		Wood rays					
Type	Diameter (μm)	Kind	Arrangement	Seriation	Composition	Average height (μm)	Average width (μm)
11	12	13	14	15	16	17	18
Libriform	16–30	Simple	Unstoried	3–5	Homogeneous	532	57
Libriform	14–26	Simple	Unstoried	1	Homogeneous to heterogeneous	186	13
Libriform	15–35	1. Simple 2. Aggregate	Unstoried	1–several	Homogeneous	266 1000+	9 155
Libriform	10–25	Simple	Unstoried	1–6	Homogeneous	220	60
Libriform	10–20	1. Simple 2. Aggregate	Unstoried	1–several	Homogeneous to heterogeneous	245 480	13 90

Contd.

Table 3.10

1	2	3	4	5	6	7	8	9	10
Castanea sativa	Ring porous	Solitary or in bands	215–330	Wanting	Orbicular to elliptical	Simple	Wanting	Paratracheal and metatracheal diffuse	Wanting
Catalpa bignonoides	Ring porous	Solitary to bands of 3–5 pores	160–210	Present in summer wood vessels	Orbicular	Simple	Wanting	Paratracheal	Wanting
Celtis australis	Ring porous	Solitary or in multiples of up to 5	180–275	Occasionally present	Orbicular to angular	Simple	Tyloses occasionally present	Paratracheal and metatracheal diffuse	Wanting
Crataegus oxyacantha	Ring porous	Solitary or in pairs	20–50	Wanting	Oval to round	Simple	Tyloses present	Metatracheal diffuse	Wanting
Fraxinus excelsior	Ring porous	Solitary and in multiples of 2–3	80–240	Wanting	Orbicular to short oval	Simple	Wanting	Paratracheal	Wanting
Juglans nigra	Semi-ring porous	Solitary or in rows of up to 3	100–270	Wanting	Orbicular or angular	Simple	Tyloses fairly abundant	Metatracheal	Crystalliferous
J. regia	Semi-ring porous	Solitary or in radial rows of 2 or 3	160–250	Wanting	Orbicular, oval or angular	Simple	Tyloses present	Metatracheal	Wanting

(Contd.)

11	12	13	14	15	16	17	18
Libri-form	16–30	Simple	Unsto-ried	1	Homoge-neous	220	11
Libri-form	20–50	Simple	Unsto-ried	2–6	Homo-geneous to hetroge-neous	266	33
Libri-form	14–22	Simple	Unsto-ried	3–8	Hetero-geneous to homoge-neous	305	32
Libri-form	6–22	Simple	Unsto-ried	2–5	Hetero-geneous	105	32
Libri-form	12–22	Simple	Unsto-ried	2–6	Homo-geneous	165	62
Libri-form	15–40	Simple	Unsto-ried	2–5	Homo-geneous to hetero-geneous	225	45
Libri-form	20–40	Simple	Unsto-ried	2–4	Homo-geneous to hetero-geneous	249	43

Contd.

Table 3.10

1	2	3	4	5	6	7	8	9	10
Morus alba	Ring porous	Solitary, concentric uninterrupted band	40–250	Present in summer wood vessels	Orbicular, oval to angular	Simple	With depositions	Paratracheal	Wanting
M. nigra	Ring porous	Solitary or in concentric bands	60–250	Occasionally present	Orbicular to oval or angular	Simple	Occasionally with depositions	Paratracheal confluent	Wanting
Parrotiopsis jacquemontiana	Diffuse porous	Solitary or in pairs	20–60	Wanting	Oval to orbicular	Simple	Wanting	Metatracheal diffuse	Wanting
Platanus orientalis	Diffuse porous	Solitary or in multiples of 2–5	40–100	Wanting	Oval to orbicular	Simple and scalariform	Occasionally occluded with tyloses	Paratracheal and metatracheal diffuse	Wanting
Populus alba	Semiring to diffuse porous	Solitary or in multiples of 2 or more	75–150	Wanting	Orbicular to oval or angular	Simple	Wanting	Terminal	Wanting
P. euphratica	Semiring to diffuse porous	Solitary or in multiples of 2 or more	70–150	Wanting	Oval to angular	Simple	Wanting	Terminal	Wanting
P. nigra	Semiring to diffuse porous	Solitary or in multiples of 2 or more occasionally crowded	65–165	Wanting	Orbicular to oval or angular	Simple	Wanting	Terminal	Wanting

(contd).

11	12	13	14	15	16	17	18
Libriform gelatinous	18–26	Simple	Unstoried	3–7	Homogeneous to heterogeneous	445	53
Libriform	12–14	Simple	Unstoried	3–7	Homogeneous to heterogeneous	380	46
Libriform	8–24	Simple	Unstoried	2–4	Heterogeneous	195	26
Libriform	20–45	Simple	Unstoried	3–14	Homogeneous	290	106
Libriform	20–40	Simple	Unstoried	1	Essentially homogeneous	195	11
Libriform	18–35	Simple	Unstoried	1	Essentially homogeneous	210	10
Libriform	18–32	Simple	Unstoried	1	Essentially homogeneous	205	10

Contd.

Table 3.10

1	2	3	4	5	6	7	8	9	10
Prunus amygdalus	Ring porous	Solitary or in multiples of 2 or more	50–110	Occasionally present	Orbicular to oval	Simple	Wanting	Paratracheal and metatracheal	Wanting
P. armeniaca	Semi-ring to ring porous	Solitary or in multiples of 2–5	40–120	Wanting	Orbicular to oval or angular	Simple	Wanting	Paratracheal	Wanting
P. domestica	Semi-ring porous	Solitary or in multiples of 2–5	50–90	Present	Broad oval to orbicular	Simple	Gummy infiltrations present	Sparse paratracheal	Wanting
P. malus (*Malus sylvestris*)	Semi-ring to ring porous	Solitary or in multiples of 2 or more	60–125	Occasionally present	Oval to orbicular or angular	Simple	Wanting	Sparse paratracheal and metatracheal	Wanting
P. persica	Ring porous	Solitary or in multiples of 2–6	45–120	Wanting	Orbicular to oval	Simple	Wanting	Sparse paratracheal	Wanting
Pyrus communis	Semi-ring porous	Solitary or in multiples of 2–6	40–125	Wanting	Oval to orbicular	Simple	Wanting	Metatracheal diffuse	Wanting

(contd).

11	12	13	14	15	16	17	18
Libriform	12–26	Simple	Unstoried	2–5	Homogeneous to heterogeneous	245	43
Libriform	13–30	Simple	Unstoried	2–5	Homogeneous to heterogeneous	280	39
Libriform	18–24	Simple	Unstoried	1–6	Homogeneous to heterogeneous	280	42
Libriform	12–30	Simple	Unstoried	2–6	Homogeneous to heterogeneous	310	38
Libriform	15–25	Simple	Unstoried	2–7	Homogeneous to heterogeneous	296	52
Libriform	8–26	Simple	Unstoried	2–7	Essentially homogeneous	210	32

Contd.

Table 3.10

1	2	3	4	5	6	7	8	9	10
P. pashia	Semi-ring porous	Solitary or in multiples of 2–5	50–95	Wanting	Oval to orbicular	Simple	Wanting	Metatracheal diffuse	Wanting
Quercous robur	Ring porous	Solitary or in bands of up to 4	180–380	Wantting	Orbicular to oval	Simple	Wantting	Paratracheal and metatracheal	Wanting
Quercous semicarpifolia	Ring porous	Bands of 1–4	200–430	Wanting	Orbicular to oval	Simple	Wanting	Paratracheal and metatracheal	Wanting
Robinia pseudoacacia	Ring porous	Solitary in multiples of 2–3 or in nest-like groups forming continuous bands	80–260	Present	Orbicular to oval or angular vestured	Simple	Tyloses present	Paratracheal	Wanting
Salix alba	Semi-ring to diffuse porous	Solitary or in multiples of up to 4	50–160	Wanting	Orbicular to oval	Simple	Wanting	Terminal	Wanting
S. wallichiana	Diffuse porous	Solitary or in pairs	30–80	Wanting	Orbicular to angular	Simple	Wanting	Terminal	Wanting

Keys and Criteria for Identification 93

(contd).

11	12	13	14	15	16	17	18
Libriform	8–20	Simple	Unstoried	2–4	Homogeneous to heterogeneous	180	25
Libriform	12–24	Oak type and simple	Unstoried	10–20	Homogeneous	800+ 160	190 9
Libriform	4–22	Oak type and simple	Unstoried	8–20+ 1	Homogeneous	800+ 173	275 8
Libriform, occasionally gelatinous	11–25	Simple	Unstoried	3–6	Homogeneous to heterogeneous	406	38
Libriform	10–25	Simple	Unstoried	1	Essentially heterogeneous	170	12
Libriform	8–20	Simple	Unstoried	1	Heterogeneous	155	9

Contd.

Table 3.10

1	2	3	4	5	6	7	8	9	10
Ulmus wallichiana	Ring porous	Wavy bands	140–180	Occasionally present	Orbicular to angular	Simple	Tyloses occasionally present	Paratracheal and metatracheal diffuse	Wanting
Viburnum sp.	Diffuse porous	Solitary or in pairs	40–75	Wanting	Orbicular to oval	Simple	Wanting	Apotracheal sparse	Wanting

(Contd).

11	12	13	14	15	16	17	18
Libriform	15–21	Simple	Unstoried	4–6	Homogeneous	380	70
Libriform	8–16	Simple	Unstoried	1–2	Heterogeneous	175	9

CHAPTER 4

ARTIFICIAL CARBONIZATION

The archaeological plant remains were, as they usually are, preserved mostly in carbonized form. The seeds, fruits, and wood pieces have been reduced to carbon while retaining, more or less, their characteristic shape. In order to ascertain the changes in shape, size, proportions, and anatomy caused by carbonization, artificial carbonization of extant grains and seeds of *Triticum monococcum, T. dicoccum, T. turgidum, T. aestivum, T. sphaerococcum, Hordeum vulgare* (both naked and hulled forms), *Oryza sativa* (two local cultivars), *Avena fatua, A. sativa, Phaseolus aureus,* and *Lens culinaris* and woods of *Pinus wallichiana* and *Robinia pseudoacacia* was carried out by baking the extant materials in an electric oven at 200°C for 12 hours.

The results (figs. 4.1–4.8) indicate that in all the five species of *Triticum* the length of the grains decreases with carbonization whereas the breadth increases. In *Triticum monococcum* the length decreases by 7.33% and the breadth increases by 18.18%. In *T. dicoccum,* decrease in length is 6.84% and increase in breadth 21.42%. Similarly, *T. turgidum* shows 16.1% decrease in length and 26.4% increase in breadth. *Triticum sphaerococcum* shows 12.08% decrease in length and 7.3% increase in breadth, while *T. aestivum* shows 18.92% decrease in length and 38.8% increase in breadth.

All the species of *Hordeum* show similar behaviour. The length decreases by 10.63% in *H. spontaneum* and breadth increases by 9.58%. In *H. distichum* length decreases by 11.7% and breadth increases by 14.2%. *Hordeum hexaploidum* shows 28.9% decrease in length and 12.6% increase in breadth. In *H. vulgare* (hulled form) the decrease in length is 7.6% and the increase in breadth is 6.59%.

Avena fatua shows 13.8% decrease in length and 21.62% increase in breadth whereas in *A. sativa* decrease in length is 7.07% and increase in breadth is 23.03%.

Decrease in length is 11.56% and 11.34% in *Oryza sativa* cultivar China 1039 and *O. sativa* cultivar Noon Beoul respectively whereas the increase in breadth is 9.6% and 9.4% respectively.

The pulses carbonized show decrease in both length and breadth. A 10.28% decrease in length and 5.55% decrease in breadth is observed in *Phaseolus aureus*. In *Lens culinaris* the decrease in length is 15.83% and the decrease in breadth 10.93%.

In all the species of *Triticum*, *Avena*, and *Hordeum* the morphological features are well preserved on carbonization. In the awned strain of rice, *Oryza sativa* cultivar Noon Beoul, the awn was lost in almost all the grains. The husk got detached in about 80% of grains of cultivar China 1039 and in about 35% of cultivar Noon Beoul. Wherever retained, morphological features are characteristically well preserved. In *Phaseolus* spp. the hilum is lost in 70% to 80% of the seeds but its position remains clear. The seed coat is partly or wholly lost in almost all the seeds of *Phaseolus* spp. and *Lens culinaris*.

The examination of woods under binocular microscope revealed that the anatomical features are well preserved in both *Pinus* and *Robinia*.

The foregoing account reveals that whereas the size varies with carbonization, the morphological and anatomical features remain well preserved.

It has been pointed out by Hopf (1955, 1957) that the prehistoric cereal grains are smaller in size than their exact present-day counterparts even when these are artificially carbonized. Our studies also confirmed these observations. Table 4.1 summarizes the measurements of modern fresh, modern carbonized, and archaeological carbonized seeds and grains.

At this stage it would be pertinent to look into the causes of carbonization of archaeobotanical material. It has been suggested that 'spontaneous combustion' (Biffen 1934), referred to as a slow process of aging (Dimbleby 1967), can occur, as when a rick of green hay ferments internally to such an extent that it builds up a high enough temperature to cause it to ignite. It has been suggested that it was through such combustion in his pile of bedding—perhaps grass or bracken—that man first discovered fire (Dimbleby 1967). However, it appears that, in nature, fire is essential for the conversion of plant material into elemental carbon (Hopf 1986).

Table 4.1. Dimensions of modern fresh, modern carbonized, and archaeological carbonized grains and seeds.

Botanical species	Nature	Length (mm)	Breadth (mm)	Source
Triticum monococcum	Modern fresh	7.5	2.2	
	Modern carbonized	6.95	2.6	
	Archaeological carbonized	5.18	2.25	Hopf (1955)
T. dicoccum	Modern fresh	7.3	2.8	
	Modern carbonized	6.8	3.4	
	Archaeological carbonized	5.71	3.06	Hopf (1955)
T. sphaerococcum	Modern fresh	4.8	2.98	
	Modern carbonized	4.22	3.2	
	Archaeological carbonized	3.7	2.7	Present study
T. aestivum	Modern fresh	6.34	3.04	
	Modern carbonized	5.14	4.22	
	Archaeological carbonized	4.37	2.51	Present study
Hordeum vulgare hulled form	Modern fresh	7.8	3.64	
	Modern carbonized	7.2	3.88	
	Archaeological carbonized	5.12	2.56	Present study
H. vulgare naked form	Modern fresh	7.3	3.2	
	Modern carbonized	6.5	3.4	
	Archaeological carbonized	5.2	2.58	Hopf (1955)

Artificial Carbonization

Oryza sativa	Modern fresh	7.7	2.95	Present study
	Modern carbonized	6.9	2.9	
			7	
	Archaeological carbonized	5.4	2.75	
Avena fatua	Modern fresh	8.1	1.5	Present study
	Modern carbonized	6.96	1.8	
	Archaeological carbonized	3.4	1.1	
A. sativa	Modern fresh	9.9	2.0	Present study
	Modern carbonized	9.2	2.5	
	Archaeological carbonized	4.5	1.7	
Phaseolus aureus	Modern fresh	4.3	3.2	Present study
	Modern carbonized	3.8	3.0	
	Archaeological carbonized	3.6	1.9	
Lens culinaris	Modern fresh	4.8	2.3	Present study
	Modern carbonized	4.04	2.3	
	Archaeological carbonized	3.95	2.1	

Palaeoethnobotany

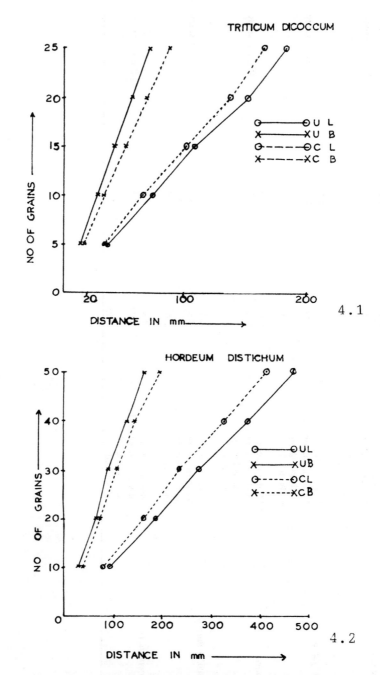

Figure 4.1-4.2: Histograms showing changes in dimensions due to artificial carbonisation of *Triticum dicoccum, Hordeum Distichum*.

Artificial Carbonization

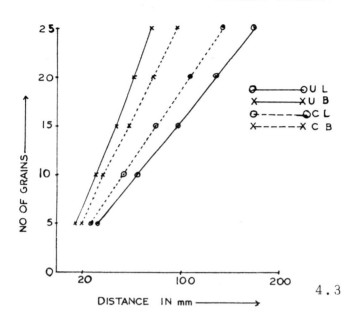

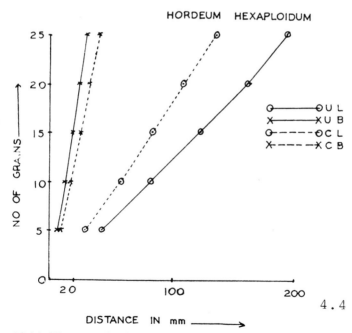

Figure 4.3-4.4: Histograms showing changes in dimensions due to artificial combination of *Triticum Turgidum, Hordeum Hexaploidum*.

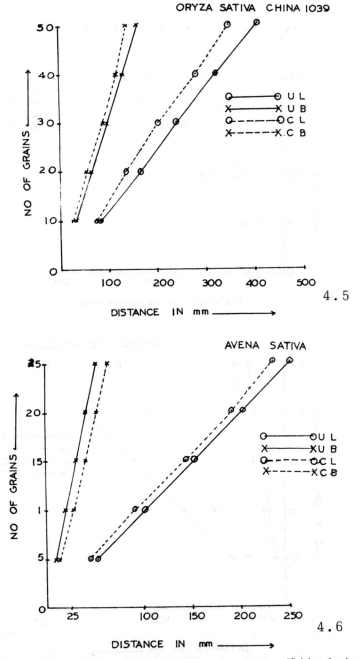

Figure 4.5-4.6: Histograms showing changes in Dimensions due to artificial corbonization of *Oryza sativa, Avena sativa*.

Artificial Carbonization

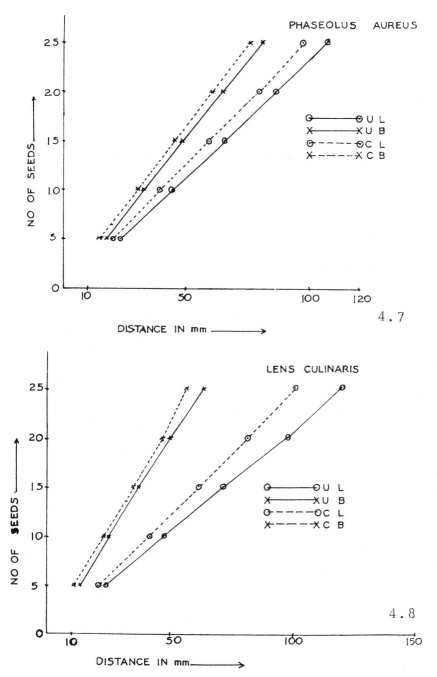

Figure 4.7-4.8: Histograms showing changes in Dimensions due to artificial carbonization of *Phaseolus aureus, Lens culinaris*.

CHAPTER 5

ARCHAEOLOGICAL EVIDENCE

5.1. CEREALS

5.1.1 *Oryza sativa* (figs. 5.1–5.4)

Caryopses referred to *Oryza sativa* were recovered from all the five phases at Semthan and from Megalithic and Post-Megalithic phases at Burzahom. Caryopses are puffed, oblong to oval in shape, 4.0 mm to 5.5 mm long and 1.7 mm to 2.7 mm broad. The grains are ribbed, the number of ribs varying from two to four (usually three). The embryo is lost but its position on the lateral side is clear in all the grains. A few grains have portions of husk (lemma and palea) well preserved at places.

The cross-sections of the caryopses do not reveal any helpful anatomical structure. Various tissues are not distinguishable because of transformation into black mass during carbonization.

Grains partially covered with husk were scanned under electron microscope. The SEM of the surface shows somewhat thick and sinous-walled cells. The SEM of the husk shows that it is composed of regular squares of characteristic chessboard pattern. At places hair bases are also visible.

Oryza sativa has been usually classified into three subspecies, namely, *indica* Kato; *japonica* Kato, and *javanica* Kato (Purseglove 1974). The distinction between the subspecies is based on morphological characters and adaptations to temperature and photoperiod conditions prevailing in different rice regions of the world (Ghose et al. 1960). The size statistics of modern rice grains and rice grains from Burzahom and Semthan is given in tables 5.1, 5.2, and 5.3 respectively. On comparison it is clear that the rice grains of the ancient material approach very close to *O. sativa* complex with respect to various statistical ratios. However, the material does not fit exactly in any one of the three subspecies *indica, japonica*, and *javanica*.

Archaeological Evidence 105

5.1   5.2

Figure 5.1: Archaeological rice caryopses from Semthan.
Figure 5.2: Impressions of rice on mud clods from Burzahom.

Figure 5.3: Chessboard pattern of Burzahom rice impressions.

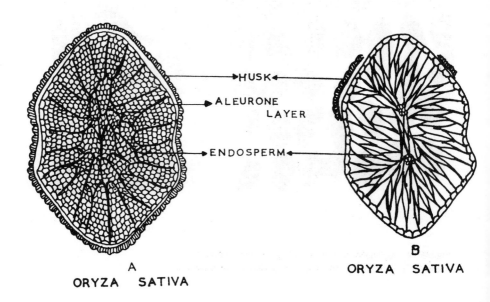

Figure 5.4: Cross-section of extant rice caryopsis (left) and archaeological rice caryopsis (right).

Hector (1937) classified *O. sativa* into four groups on the basis of size alone.
1. Long: length 4-7 times the breadth.
2. Fine: length 3-4 times the breadth.
3. Coarse: length 2-3 times the breadth.
4. Round: length 2 times the breadth.

Ghose et al. (1960) also presented a classification for *O. sativa* of Indian origin as follows:
1. Coarse: 2.75 mm and greater breadth.
2. Fine: 2–3.75 mm breadth.
3. Superfine: 2 mm and lesser breadth.

On comparison it is clear that the rice grains of the ancient material come under the Coarse group of Hector. By Ghose's classification the material comes under the Fine group.

Table 5.1. Size statistics of modern rice grains.

Species	Length	Breadth	Thickness	L/B	L/T	B/T	L/(B × T)
O. perennis	8.12	2.29	1.61	3.54	5.04	1.42	2.21
O. officinalis	4.25	2.12	0.85	2.0	5.0	2.49	2.36
O. rufipogon	7.0	2.65	1.0	2.64	7.0	2.65	2.64
O. sativa var. javanica	8.82	2.55	1.95	3.45	4.52	1.30	1.37
O. sativa var. japonica	6.50	2.75	1.38	2.36	4.71	1.27	1.71
O. sativa var. indica	5.25	1.88	1.22	2.79	4.30	1.54	1.70

Table 5.2. Size statistics of rice grains from Burzahom.

	Average	Maximum	Minimum
Length (L) in mm	4.5	5.2	4.0
Breadth (B) in mm	2.0	2.5	1.7
Thickness (T) in mm	1.2	1.7	1.1
L/B	2.25	2.08	2.35
L/T	3.75	3.05	3.63
B/T	1.66	1.47	1.54
T/B	0.6	0.68	0.64
L/(B × T)	1.87	1.22	2.13

Table 5.3. Statistics of rice grains from Semthan.

	Average	Minimum	Maximum
Length (L) in mm	4.7	4.3	5.5
Breadth (B) in mm	2.1	1.9	2.7
Thickness (T) in mm	1.4	1.2	1.8
L/B	2.2	2.3	2.1
L/T	3.5	3.58	3.06
B/T	1.5	1.58	1.5
T/B	0.66	0.63	0.66
L/ (B × T)	1.60	1.88	1.13

COMMENTS

The genus *Oryza* consists of 20 wild species (both diploid and tetraploid forms) and two cultigens, namely, *O. sativa* L. and *O. glaberrima* Steud (Chang 1986a). Among the wild relatives the perennial *O. rufipogon* Griff. syn. *O. perennis* Moench is generally considered to be the ancestor of *O. sativa* (Chang 1976a, 1976b, 1976c, 1983, 1985a, 1985b, 1986a, 1986b, Second 1982). On the other hand, *O. glaberrima*, whose economic importance is restricted to west and central Africa, is closely related to the African species, *O. breviligulata* A. Chev et Roehr syn. *O. barthii* A. Chev., but their genetic affinity has been differentially interpreted. According to Porteres (1950), Morishima et al. (1963), Oka (1974), and Chang (1976b), *O. glaberrima* was domesticated from *O. breviligulata*. Others contend that *O. glaberrima* was introduced into Africa from Asia and *O. breviligulata* was a hybrid derivative either between *O. glaberrima* and *O. longistaminata* A. Chev et Roehr (Richaria 1960) or between *O. glaberrima* and a more recent introduction of *O. satvia* in Africa (Nayar 1973). Some of the phylogenetic relationships proposed are presented in text fig. 1. According to Oka (1974), *O. sativa* may have been domesticated independently in various areas. Its differentiation is under selective forces in the course of domestication. Chang (1976b, 1986a) believes that the present-day distribution of wild rice is an outcome of the continental drift. The Gondwanaland origin of rice and its interspecific differentiation before the supercontinent fractured and drifted apart are indicated by the pantropical distribution of the wild species of the genus in a non-disjunct manner across Africa, Asia, Oceania, and Latin America. Nayar (1973) proposed that *O. glaberrima* was introduced in Africa from Asia, and *O. breviligulata* escaped from cultivated fields following the introgression of genes of *O. sativa* in *O. glaberrima*. Recently Second (1982) proposed three primary domestications. The differentiation of cultivated rice resulted in part from the geographical differentiation of races of *O. rufipogon*. Some of the weedy forms of *O. breviligulata* evolved through introgression of genes of *O. sativa* into *O. glaberrima*.

The chronology of rice cultivation in different parts of Asia as revealed by archaeobotanical finds is summarized in table 5.4 (Chang 1986a, Lone et al. 1986b). The wild variety of rice is reported from protoneolithic levels at Chopane Mando (ninth to eighth millennium B.C.) in the Belan valley of the Vindhya ranges (Sharma and Misra 1980). The oldest record of cultivated rice is from Mahagara and Koldihawa in the same region. The dates tend to support the view that the Asian cultigen of rice evolved over a broad belt that extended from the southern foothills of the Himalayas, across upper Burma, northern Thailand, and Laos, to north Vietnam and southwest and south China (Chang 1976b, 1986a, Glover 1977). The chronology avail-

Archaeological Evidence

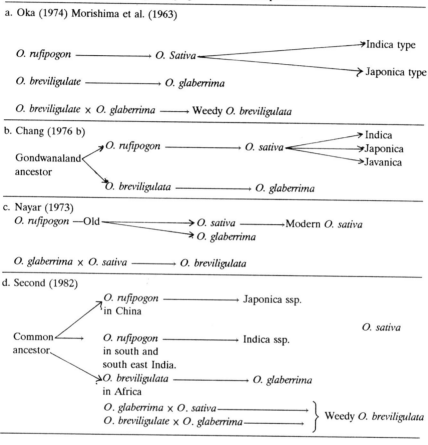

Textfig. 1 Phylogenetic relationships in rice.

able till date reveals that the Vindhyan region might have the credit for originating the domestication of *O. sativa*.

Oryza glaberrima has its primary centre of diversity in the swampy areas of the Upper Niger River and its two secondary centres are to the southwest near the Guinean coast. The primary centre was formed c. 1500 B.C. and the secondary centres around 1000 B.C. (Porteres 1950, Chang 1975).

So far as the Kashmir valley is concerned, the excavations at Burzahom (2325 B.C.) and Gofkral (2100 B.C.) do not reveal rice in the early Neolithic levels (Sharma 1982, Buth et al. 1986a). Thus the rice finding at Semthan (c.1500 B.C.) is very interesting. Rice, which is the main crop of the valley today, appears at Gofkral toward the end of the Neolithic II period datable to c. 1000 B.C. (Sharma 1982) and in the Megalithic period at Burzahom. It becomes evident that the rice culture was introduced into the valley somewhere around 1500 to 1000 B.C. At that time rice culture was well established in the Indo-Gangetic plains and apparently rice was introduced from there along with the migration of people. The people of Neolithic Kashmir were in definite contact with those of the Indo-Gangetic plains at that time (Buth and Kaw 1985, Buth et al. 1986a).

5.1.2. *Triticum aestivum* (figs. 5.5–5.7)

The evidence of *Triticum aestivum* as a food plant comes from all the phases at both sites. Caryopses are oval to subglobular, rather plump, varying in length from 3.7 mm to 5.2 mm and in breadth from 2 mm to 3.8 mm. The two cheeks of the grains are fairly wide and flat. The embryo or the position of the embryo is at the base of the dorsal surface. The dorsal surface is raised and a fairly deep furrow is present on the ventral side. The caryopses are naked and some of them are partially or completely covered with a thin layer of pericarp.

The cross-sections of the caryopses do not reveal much useful anatomical data. After carbonization the cells of the caryopses inside have been distorted. The black mass formed shows little detail of the structure of individual cells. The only preserved tissue appears to be endosperm and the remains of vascular tissue in the furrow.

The SEM of the surface reveals a characteristic cell alignment and relief. The pericarp is made up of horizontally placed parenchymatous cells.

Palaeoethnobotanists have emphasized the importance of size of the grains in determining the identity of archaeological cereals. Therefore we attempted to determine the size statistics of extant and archaeological wheats. Size statistics of some extant hexaploid wheat species and of wheat grains from Burzahom and Semthan are given in tables 5.5, 5.6, and 5.7. Matching the different indices, the ancient wheat comes very close to *T. aestivum* and *T. compactum* while differing considerably from *T. sphaerococcum*.

Hexaploid wheats have further been classified by Mackay (1954) and Sears (1956). Taking Mackay's classification, the species *T. aestivum* (L.) em Thell. includes subspecies *T. spelta* (L.) Thell, *T. macha* (Dek et Men) Mackay, *T. vavilovii* (Tuman) Sears, *T. vulgare* (Vill) Host Mackay, and *T. sphaerococcum* (Perc) Mackay. While working on archaeological material of naked wheats, Helbaek (1966b, 1969) included only *T. vulgare*, *T. compactum*, and *T. sphaerococcum* under *T. aestivum*. In view of this, the present archaeological material is placed under *T. aestivum* (L.) em Thell complex.

5.1.3. *Triticum sphaerococcum* (figs. 5.8 and 5.9)

Besides the wheat grains referred to *T. aestivum* above, a total of 50 grains

Archaeological Evidence 111

Figure 5.5: Archaeological caryopsis of *Triticum aestivum* from Semthan.

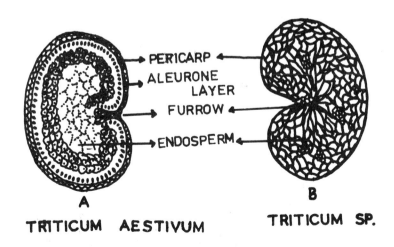

Figure 5.6: Cross-section of extant (left) and archaeological (right) *Triticum aestivum* caryopses.

Table 5.4. Chronology of some oldest rice remains from Asia

Site	Type of plant remains	Estimated age	Reference
Koldihwa and Mahagara, U.P., India	Rice grains embedded in earthen potsherds and husks in cow dung	6570–4530 B.C.	Sharma and Mandal (1980) Sahara and Sato (1984)
Chirand, Bihar, India	Charred grains	2500–1800 B.C.	Vishnu-Mittre (1972a)
Atranjikhera U.P., India	Charred grains	2,000–60 B.C.	Chowdhury et al. (1977)
Rangpur, Saurashtra, India	Impressions	2,000–1500 B.C.	Vishnu-Mittre and Savithri (1982)
Lothal, Gujarat, India	Impressions	2300–1700 B.C.	Vishnu-Mittre (1961) Vishnu-Mittre and Savithri (1982)
Ahar, Rajasthan, India	Impressions	1885–1070 B.C.	Vishnu-Mittre (1969)
Semthan, Kashmir, India	Charred grains	c. 1500 B.C.	Present study
Ben Chiang, Thailand Ban Na Di, Thailand	Husk remains in potsherds, Hulled rice kernels	3500 B.C. 1500–900 B.C.	Yen (1982) Higham and Kijngam (1984) Chang and Loresto (1984)
Ulu Leang, S. Sulawesi, Indonesia.	Carbonized grains and glume fragments	c. 4000 B.C.	Glover (1977)

Solana, N. Luzon, Philippines	Glume imprints on potsherds	c. 1400 B.C.	Chang (1986a)
Luo-Jia Jiao, Zhejiang, China	Carbonized grains and broken husks	c. 5000 B.C.	Team of Luo-Jia-Jiao Site (1981)
Ho-mu-tu, Zhejiang, China	Carbonized grains, husks straw	5000 B.C.	Chekiang Provincial Cultural Commission and Chekiang Provincial Museum 1976 Li 1983
Heng Chiun, Taiwan	Glume imprints on potsherds	1985 B.C.	Wang 1984
Chih-Shan-Yen, Taiwan	Carbonized grains and brown rice	2095–1500 B.C.	Chang 1983
Yang-Shao, Honan China	Glume imprints on pottery	c. 3200–2500 B.C.	Dao 1985
Xom Trai Cave, Vietnam		4000–2000 B.C.	

Figure 5.7: SEM of archaeological *Triticum aestivum* 500x.

from Burzahom (all the four phases) and 14 grains from Semthan (Pre-N.B.P. and Kushan phases) were recovered which no doubt belong to wheats but are certainly not those of *T. aestivum*. These caryopses are oval to subglobular, comparatively short and rounded, and rather plump when viewed from the ventral side. They vary in length from 3 mm to 4.7 mm and in breadth from 2.2 mm to 2.5 mm. The SEM of the surface shows resemblance in relief, cell pattern, and cell alignment to that of *T. sphaerococcum*.

Table 5.5. Size statistics of modern hexaploid wheat grains

Species	L/B	L/T	B/L	T/L	T/B	L/(B x T)
T. aestivum	2.68	3.07	0.37	0.32	0.87	1.32
T. compactum	2.44	3.07	0.40	0.32	0.79	1.26
T. sphaerococcum	1.76	1.62	0.56	0.61	1.09	0.65

The size statistics of these caryopses is given in tables 5.8 and 5.9. Comparing the various indices with those of extant wheat species it is seen that the ancient caryopses approach very close to *T. sphaerococcum*. Hence these caryopses are placed under *T. sphaerococcum* (Perc.) Mackay.

Archaeological Evidence 115

Figure 5.8: Archaeological *T. sphaerococcum* caryopses from Burzahom.

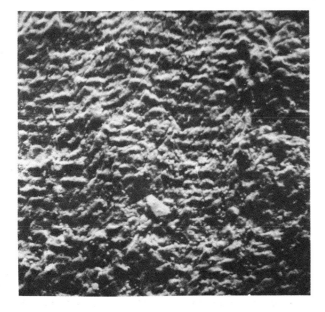

Figure 5.9: SEM of archaeological *T. sphaerococcum* caryopsis 200x.

Table 5.6. Size statistics of wheat *Triticum aestivum* grains from Burzahom

	Average	Maximum	Minimum
Length (L) in mm	4.9	3.8	5.2
Breadth (B) in mm	2.5	2.1	3.8
Thickness (T) in mm	2.0	1.7	2.4
L/B	1.96	1.80	1.36
L/T	2.45	2.23	2.16
B/L	0.51	0.55	0.73
T/L	0.40	0.44	0.46
T/B	0.8	0.80	0.63
L/(B × T)	0.98	1.06	0.57

Table 5.7. Size statistics of wheat (*T. aestivum*) grains from Semthan.

	Average	Maximum	Minimum
Length (L) in mm	4.8	3.7	5.2
Breadth (B) in mm	2.4	2.4	3.5
Thickness (T) in mm	1.9	1.6	2.4
L/B	2.0	1.85	1.48
L/T	2.52	2.31	2.16
B/L	0.5	0.54	0.67
T/L	0.20	0.43	0.46
T/B	0.79	0.8	0.68
L/(B × T)	1.05	1.15	0.61

Table 5.8. Size statistics of wheat *T. sphaerococcum* grains from Burzahom.

	Average	Maximum	Minimum
Length (L) in mm	4.0	3.0	4.5
Breadth (B) in mm	3.2	2.8	3.5
Thickness (T) in mm	2.8	2.2	3.0
L/B	1.25	1.07	1.28
L/T	1.42	1.36	1.5
B/L	0.8	0.93	0.77
T/L	0.7	0.73	0.66
T/B	0.87	0.78	0.85
L/(B × T)	0.44	0.48	0.42

COMMENTS

Sakamura reported as early as 1918 that wheats form a ployploid series with 14, 28 and 42 chromosomes. Later, Kihara (1924) recognized three types of genomes, each composed of seven chromosomes, and labelled them A, B, and D. The diploid wheats have only the A genome, tetraploids A and B, and the hexaploids A, B, and D. Further studies revealed the possible

Table 5.9. Size statistics of wheat *T. sphaerococcum* from Semthan.

	Average	Maximum	Minimum
Length (L) in mm	3.7	3.0	4.7
Breadth (B) in mm	2.7	2.2	3.3
Thickness (T) in mm	2.6	2.1	3.0
L/B	1.37	1.36	1.42
L/T	1.43	1.43	1.56
B/L	0.73	0.74	0.70
T/L	0.70	0.70	0.69
T/B	0.96	0.95	0.90
L/(B × T)	0.53	0.65	0.47

sources of A, B, and D genomes and models for the evolution of diploid, tetraploid, and hexaploid wheats were proposed (Kihara 1944, McFadden and Sears 1946, Peterson 1965, Riley 1965, Feldman 1976).

While the diploid wheats have only the A genome, the sources of B and D genomes appear to be in the closely related genus *Aegilops*. The B genome is believed to have been donated by an ancestor of the present-day *Aegilops speltoides*, whose genomes SS appear to be closely similar to the BB genomes of tetraploid wheat (Peterson 1965). Similarly, *Aegilops squarrosa* is suggested as the donor of D genome (Kihara 1944, McFadden and Sears 1946). The possible evolution of cultivated diploid, tetraploid, and hexaploid wheats is presented in text figs. 2 and 3.

Distribution of wild relatives of wheat as well as distribution, ecological behavious, and genetic interaction of the weed races and archaeological findings indicate that the Near East, comprising the early farming villages excavated in the arc of hilly flanks from the Deh Luran Plain in Iran through south east Turkey to southern Jordan, is the centre of origin and domestication of this crop as early as 7000 B.C. (Harlan and Zohary 1966, Harlan 1971). The diploid wild einkorn *Triticum boeoticum* has been found at Ali Kosh, in the Bus Mordeh Phase c. 7500–6750 B.C. (Helbaek 1966b); at Jarmo c. 6750 B.C. (Helbaek 1959b); at aceramic Hacilar c. 7000 B.C. (Helbaek 1966b); and at Tell Mureybit c. 8050–7542 B.C. (Van Zeist and Casparie 1968).

Cultivated einkorn *T. monococcum* is recorded at Ali Kosh, Bus Mordeh Phase 7500–6750 B.C. (Helbaek 1966b); Tell es Sawwan 5800–5600 B.C. (Helbaek 1965a); Jarmo c. 6750 B.C. (Helbaek 1959b); Catal Huyuk 5850–5600 B.C. Helbaek (1964a); late Neolithic Hacilar (Helbaek 1966b); Ghediki 6000–5000 B.C. (Renfrew 1965); Argissa 6000–5000 B.C. (Hopf 1962); Azmaska c. 5000 B.C; and Karanova c. 5000 B.C; (Hopf 1957 cf. Renfrew 1969). Consequently Helbaek (1966b) postulated that 'West Central Anatolia was the primary centre of conscious development and selection took place about 6000 B.C.'

The tetraploid wheat *T. dicoccoides* (wild emmer) has been identified from Jarmo c. 6750 B.C. only (Helbaek 1960b, 1966b), whereas cultivated *T. dicoccum* is found in all the earliest sites of the Near East, Anatolia, and Southern Europe (Helbaek 1960a, Renfrew 1973). The other tetraploid wheats are rather scanty in the archaeological contexts as at Fayum, Egypt, fifth millennium B.C., *T. durum* is reported (Tackholm et al. 1941).

Similarly, at Beycesultan, *T. durum* and *T. turgidum* are suspected, and in Spain, *T. turgidum* is suspected (Hopf 1970). Buth (1970) reported *Triticum* spp. (cf. *T. dicoccum*) from Nubia Egypt 2500 B.C. Noy et al. (1975) have provided the evidence of emmer as early as 13800-14800 B.C. at Nehal Oren (Kebaran).

The present evidence reveals that hexaploid wheats *T. aestivum* and *T. compactum* are of equal antiquity (about 5500 B.C. Renfrew 1973). In the Near East *T. aestivum* is found at Tape Sabz 5500-5600 B.C. (Helbaek 1966b); Tell es Sawwan c. 5800-5600 B.C. (Helbaek 1965a); Catal Huyuk 5850-5600 B.C. (Helbaek 1964a, 1966b); and late Neolithic Hacilar 5800-5000 B.C. (Helbaek 1966b). *Triticum compactum* has been recorded in the pre-pottery Neolithic B levels at Tell Ramad (Van Zeist and Botteima 1966) and many other sites (Zohary and Hopf 1986).

On spelt wheat *T. spelta*, Helbaek (1966b) holds the view that it 'never occurred in prehistoric west Asia and even in Europe which seems to be its area of origin (as a cultivar at least), it is rather a newcomer. The remaining hexaploid wheat, *T. sphaerococcum*, is found from the sites in northwest India dating to the third millennium B.C. It is recorded at Harappa 2250 B.C. (Burt 1941); Mohenjodaro 2250 B.C. (Stapf 1931, Shaw 1943); Chandudaro (Shaw 1943); and Burzahom and Semthan (present study). In view of these finds and its absence from the sites in the Near East and Europe, *T. sphaerococcum* is supposed to have originated in the northwestern area of the Indian subcontinent (Rao 1974). *Aegilops tauschii*, growing wild in Kashmir, might be one of its ancestors. Singh (1946) opined that due to its high resistance to drought *T. sphaerococcum* was particularly selected by our ancestors and according to Ellerton (1939) it appears to be a derivative of *T. aestivum*.

Chronology of the oldest wheat remains from India is summarized in table 5.10. It becomes evident that India has received only hexaploid naked wheats, *T. aestivum*, *T. compactum*, and *T. sphaerococcum*. Chowdhury et al. (1977) have pointed out that 'all these species have been named only recently based on conventional method of classification. In the early stage of their evolution when man used these cereals he was not in a position to separate them by the look of the plants or by other criteria. . . . When these early hexaploid wheats were introduced in north India, they were still at a stage of rapid internal changes and at the same time making an effort to adjust themselves in the new environments. As a result of all these, we

Archaeological Evidence

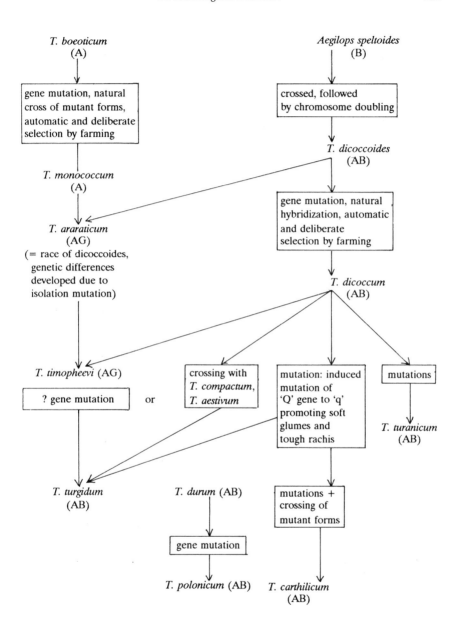

Text fig. 2: The possible evolution of cultivated diploid and tetraploid wheats (based on Peterson 1965, with genomes added after Bell in Hutchinson 1965).

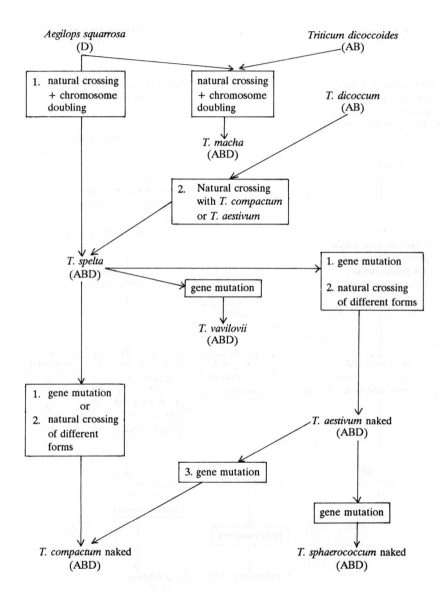

Text fig. 3: Possible evolution of hexaploid wheats (after Peterson 1965 with genomes added after Bell in Hutchinson 1965).

Table 5.10. Chronology of some oldest wheats from India

Site	Species	Estimated age	Reference
Burzahom	T. aestivum	2325 B.C.	Buth and Kaw 1985,
Kashmir	J. sphaerococcum		Present study
Mohenjodaro	T. sphaerococcum, T. compactum	2250–1750 B.C.	Stapf 1931
Harappa	T. sphaerococcum	2250–1750 B.C.	Burt 1941
Chanudaro	T. sphaerococcum, T. compactum	2250–1750 B.C.	Shaw 1943
Banawali, Haryana	T. aestivum	c. 2300 B.C.	Lone et al. 1987
Kalibangan, Rajasthan	Triticum spp.	c. 2300–1750 B.C.	Vishnu-Mittre and Savithri 1975
Gofkral, Kashmir	Triticum spp.	2100 B.C.	Sharma 1982
Chirand, Bihar	T. sphaerococcum	1800 B.C.	Vishnu-Mittre 1972a
Navdatoli	T. compactum, T. vulgare	1600–1440 B.C.	Vishnu-Mittre 1961
Semthan, Kashmir	T. aestivum, T. sphaerococcum	1500 B.C.	Present study
Atranjikhera, U.P.	T. compactum	1200–600 B.C.	Buth and Chowdhury 1973
Inamgaon, Maharashtra	T. compactum	1300–700 B.C.	Kajale 1977a

find in the early hexaploid wheats grown in India, a great variation in their morphological features'.

Recent excavations at Mehrgarh in Pakistan revealed cultivation of *T. monococcum*, *T. dicoccum*, and *T. durum* (or *T. aestivum*) in the Neolithic period I: 6000–5000 B.C. (Jarriage and Meadow 1980). This is the first and very interesting record of diploid and tetraploid wheats from the Indian region. Thus Mehrgarh offers proof of the earliest centre of cereal cultivation in the Indo-Pak region. A rethinking of the belief that India received only hexaploid wheats is needed. It may be that the Indo-Pak region was an independent or secondary centre of origin and domestication of wheats.

In the Kashmir valley the hexaploid wheats *T. sphaerococcum* and *T. aestivum* as revealed by the present study date back to about 2325 B.C., which corresponds to the Harappan age. Thus the valley appears to have received wheats as early as the Harappans and other contemporary cultures of India and the wheats have been continuously in use from the time of their domestication here.

5.1.4. *Hordeum vulgare* (figs. 5.10 and 5.11)

Hordeum vulgare has been recovered from all the phases of both sites. Caryopses are 5 mm to 6.5 mm long and 2 mm to 3.5 mm broad. Some of the caryopses have a distinct bulge in the centre whereas a few have a prominent twist at the anterior end. The grains are flat on the dorsal side and somewhat pointed at both ends. A shallow dorsal furrow and a deeper ventral furrow are prominent. The embryo or the position of the embryo is at the base of the dorsal surface in all the caryopses. The embryo has a pointed beak. All the caryopses are partially or completely enclosed in a thick covering which shows longitudinal ridges under binocular microscope.

The serial sections of the caryopses show nothing but a black mass of tissue. However, the furrow, low-domed lobes, and remains of vascular tissue are visible.

The caryopses were scanned under an electron microscope at various places. In SEMs the surface shows a characteristic cell alignment and relief. The pericarp consists of a layer of somewhat thick, sinous-walled parenchymatous cells. Short cells and silica bodies are also seen.

The size statistics of extant and ancient *H. vulgare* is given in tables 5.11–5.13.

COMMENTS

Barley is a self-pollinating diploid with $2n = 14$ (Takahashi 1955; Nilan 1971). The wild two-row form *Hordeum spontaneum* is believed to be the

Archaeological Evidence 123

Figure 5.10: Archaeological *Hordeum vulgare* from Semthan.

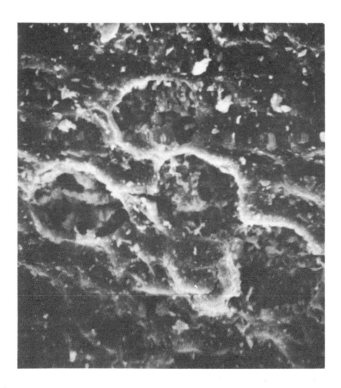

Figure 5.11: SEM of inner layer of *Hondeum vulgare* 1000x.

ancestor of all the cultivated barleys (Harlan 1976). Under domestication six-rowed races appeared as shown in text fig. 4.

Table 5.11. Size statistics of barley grains from Burzahom.

	Average	Minimum	Maximum
Length (L) in mm	5.7	4.8	6.5
Breadth (B) in mm	2.7	2.0	3.4
Thickness (T) in mm	2.3	2.1	2.5
L/B	2.1	2.4	1.91
L/T	2.47	2.28	2.6
B/L	0.47	0.41	0.52
T/L	0.40	0.44	0.38
T/B	0.85	1.05	0.73
L/(B × T)	0.91	1.14	0.76

Table 5.12. Size statistics of barley grains from Semthan.

Length (L) mm	5.8	5.0	6.5
Breadth (B) mm	2.8	2.0	3.5
Thickness (T) mm	2.1	1.55	2.25
L/B	2.07	2.5	1.85
L/T	2.76	3.2	2.9
B/L	0.48	0.4	0.53
T/L	0.36	0.31	0.34
T/B	0.75	0.77	0.64
L/(B × T)	0.95	1.6	0.83

Table 5.13. Size statistics of extant barleys

Species	L/B	L/T	B/L	T/L	T/B	L/(B × T)
H. vulgare hulled	2.14	2.43	0.46	0.41	0.87	0.67
H. vulgare haked	2.28	2.60	0.43	0.38	0.87	0.87
H. hexaploidum	5.4	5.5	0.18	0.18	0.98	3.9

So far as archaeobotanical finds are concerned several grains of barley were reported from a late Palaeolithic site, dating between 17,000 and 18,000 years B.P., in Wadi Kubbaniya near Aswan in Egypt (Wendorf et al. 1979, Stemler and Falk 1980, 1984), which of course was a tantalizing discovery. However, recent studies indicated that these grains were not associated with the late Palaeolithic occupations. The age determination of the actual cereal grains using linear accelerator radiocarbon counter revealed an age of 4850 to 820 years, indicating that these were contaminants into the late Palaeolithic levels (Wendorf and Schild 1984).

The undoubted palaeobotanical finds tend to support the belief that

Archaeological Evidence

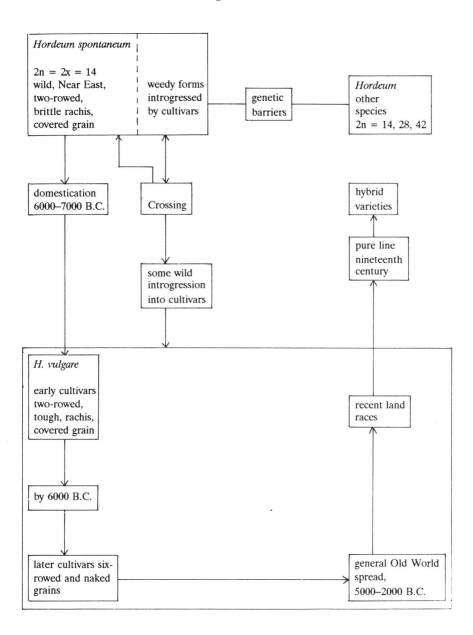

Text fig. 4: Evolution of cultivated barley *Hordeum vulgare*, after Harlan (1976).

H. spontaneum is the ancestor as it is the only form of wild barley yet found in the early farming villages (Harlan 1968, Renfrew 1969, 1973). It has been reported from Ali Kosh, Bus Mordeh phase 7500–6750 B.C. (Helbaek 1966b); Tape Guran 6200–5500 B.C. (Meldgaard et al. 1963); Jarmo 6750 B.C. (Helbaek 1959a); Tell Mureybit 8050–7542 B.C. (Van Zeist and Casperi 1968); and Beidha c. 7000 B.C. (Helbaek 1966b). The Beidha finds are of particular interest since many thousands of impressions of *H. spontaneum* were found with grains larger than in the truly wild forms and so were described as 'cultivated wild barley'. Some signs of domestication are also seen in the Jarmo finds (Helbaek 1960a).

Fully domesticated *H. distichum* occurs at many sites at a slightly later date, as in Ali Kosh 6750–5600 B.C. (Helbaek 1966b); Tape Sabz 5500–5000 B.C. (Hole and Flannery 1967); Tape Guran 6200–5500 B.C. (Meldgaard et al. 1963); Tell es Sawwan 5800–5600 B.C. (Helbaek 1965a, 1965b); Matarrah 5500 B.C. (Helbaek 1966a); and late Neolithic Hacilar 5800–500 B.C. (Helbaek 1961).

A cultivated form of *H. vulgare* has been identified at Ali Kosh 6750–6000 B.C. (Helbaek 1966b); Tape Sabz 5500–5000 B.C. (Helbaek 1966b); ceramic Hacilar c. 7000 B.C. (Helbaek 1966b); Beidha 7000 B.C. (Helbaek 1966b); Catal Huyuk 5850–5600 B.C. (Helbaek 1964a, 1964b, 1966b); Tell es Sawwan 5800–5600 B.C. (Helbaek 1965a); Mersin 5750 B.C. (Helbaek 1959b); late Neolithic Hacilar 5800–5000 B.C. (Helbaek 1961); Can Hasan c. 5250 B.C. (Renfrew 1968); and Argissa Maghula 6000–5000 B.C. (Hopf 1962).

From the earliest contexts of the Indian subcontinent, Jarriage and Meadow (1980) reported cultivation of two-row hulled *H. distichum* and naked barley *H. vulgare* var. *nudum* from Mehrgarh period I, 6000–5000 B.C. Yet another excavation at Mahagara in the Ganga basin has yielded the evidence of barley cultivation dating from the sixth to the seventh millennium B.C. (Sen Gupta 1985). Some other oldest records include *H. vulgare* var. *nudum* at Mohenjodaro 2250–1750 B.C. (Luthra 1936); *H. vulgare* var. *hexastichum* at Harappa 2250 B.C. (Vats 1941); *H. vulgare* at Banawali 2300 B.C. (Lone et al. 1987); naked and hulled barley at Kalibangan 2090–2075 B.C. (Vishnu-Mittre and Savithri 1982); and *H. vulgare* from Atranjikhera 2000–2050 B.C. (Chowdhury et al. 1977) and Gofkral (Sharma 1982).

The foregoing account and the modern distribution of wild barley (Harlan and Zohary 1966) establish that barley was domesticated in the Near East. In the Indian region its cultivation dates back to the sixth to seventh millennium B.C. Barley culture in India can be followed with certainty across northern India and then southward (Sankalia et al. 1953, Raikes and Dyson 1961, Bakshi and Rana 1974, Vishnu-Mittre 1974a). Barley culture in Kashmir is as old as that of wheat and appears to have been introduced from its place of origin through the Indian plains at a very advanced stage. Only hulled barley adapted to the Indian conditions and naked barley was confined to some isolated areas like the Ladakh region of Jammu and Kashmir.

5.1.5. *Avena* spp. (figs. 5.12 and 5.13)

Oats have been recovered from N.B.P. (*Avena fatua*), Indo-Greek (*A. fatua, A. sativa*), Kushan (*A. sativa*), and Hindu Rule (*A. sativa*) phases at Semthan. The grains are slender, elongate, 3.8 mm to 5 mm long, and 1 mm to 1.8 mm broad. The dorsal surface is smooth and a shallow furrow is seen on the ventral surface. The caryopses are partially or wholly covered with a thin layer of pericarp. The position of the embryo is dorsal. The SEM of the caryopses shows that the surface is made of a thin layer of parenchymatous cells comparable in structure to that of extant *Avena* spp.

The diploid and the tetraploid oats do not have an area of distribution in Kashmir. Of the hexaploid species occur *A. fatua* and *A. sativa*. In the absence of detailed morphological characters size and shape have proven useful in the classification of oats as there is considerable difference in size of wild and cultivated oats (Renfrew 1969). The dimensions and the proportional indices of the Semthan oats and the extant *Avena* spp. are presented in tables 5.14 and 5.15. Comparing the different indices, it becomes evident that concordance exists between the living and the ancient oats but the proportional indices of the latter do not match exactly with those of any particular species. However, based on differences in size alone, the smaller ones are referred to *Avena fatua* L. and the larger ones to *A. sativa* L.

Table 5.14. Size statistics of oat grains from Semthan

	Average	Minimum	Maximum
Length (mm)	4.55	3.8	5.0
Breadth (mm)	1.5	1.0	1.8
Thickness (mm)	1.1	0.8	1.5
L/B	3.03	3.8	2.7
L/T	4.13	4.75	3.33
B/L	0.32	0.26	0.36
T/L	0.24	0.21	0.30
T/B	0.73	0.80	0.83
L/(B × T)	2.75	4.75	1.85

Table 5.15. Size statistics of extant *Avena* spp.

Species	L/B	L/T	B/L	T/L	T/B	L/(B × T)
Avena fatua	4.59	6.02	1.31	0.16	0.76	3.43
Avena sativa	4.84	5.30	1.09	0.18	0.90	2.55

Figure 5.12: Caryopses of *Avena* spp. from Semtham.

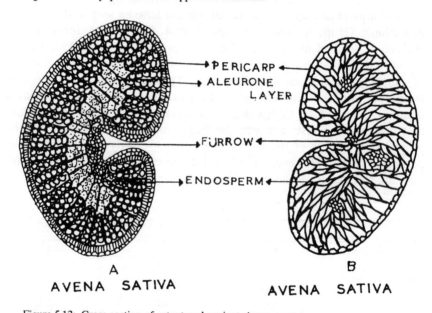

Figure 5.13: Cross-section of extant and ancient *Avena* caryopses.

COMMENTS

Avena species occur at three ploidy levels, diploids ($2n = 14$), tetraploids ($2n = 28$) and hexaploids ($2n = 42$). A summary of present knowledge on species relationships is provided in text fig 5, which is based on Holden (1966, 1976), Rajathy and Sadasivaiah (1969), Ladizinisky (1971), Ladizinisky and Zohary (1971), Zohary (1971), and Rajathy and Thomas (1974). Accordingly, either *A. magna* or *A. murphyi* must be regarded as a tetraploid ancestor of *A. sterilis*. Cytogenetic studies indicate *A. strigosa* as a genome donor to hexaploids. The occurrence of at least one other diploid species presently unknown can be predicted as a common ancestor to *A. magna, A. murphyi,* and *A. sterilis.*

Previously it was thought that *A. fatua* was the progenitor of *A. sativa* and *A. sterilis* that of *A. byzantina,* but Coffman (1946) showed that all the cultivated oats are derived from *A. sterilis.*

Finds of wild oat *A. sterilis* grains have been reported from the early Neolithic villages in the Near East at Ali Kosh 6750–5600 B.C. (Hole and Flannery 1967), Beidha 7000 B.C. (Helbaek 1966b) and Amouq A. c. 5750 B.C. (Helbaek 1960c). Only at Beidha has the species been identified as *A. ludoviciana* syn. *A. sterilis* ssp. *ludoviciana.*

Avena fatua, A. strigosa, and *A. sativa* have been reported from prehistoric contexts in Europe. The earliest oat grain from Europe comes from aceramic Neolithic levels at Achilleoin, Thessaly, Greece 6000–5000 B.C. but its species could not be identified (Renfrew 1966). *Avena fatua* has been found in the Bronze lakeside villages at Alpenquai on Lake Zurich and Morigen (Bertsch and Bertsch 1949) and in Lengyel in Hungary (Tempir 1964). In Britain the common wild oats are found at Maiden Castle, Little Salisbury, Worlebury, Meare Lake Village, Glastonbury (Helbaek 1952b), and Aldwick Barley (Renfrew 1965). *Avena strigosa* has also been reported in the Alpine region in the Bronze Age at Montellier in Savoy (Heer 1866) and in early Iron Age Britain at Maiden Castle and Fifield Bavant (Jessen and Helbaek 1944). Matthias and Schultze-Motel (1967) referred the impressions from Schraplan and Calbe to *A. sativa* on account of their large size. Surprisingly there is no record of oats from Indian archaeological excavations. Jarriage and Meadow (1980) recovered a few grains of *Avena* spp. from Mehrgarh period III, which is so far the earliest record from the Indian subcontinent. The present oat grains from Semthan have been referred to *A. fatua* as well as *A. sativa* based on the size of the grains alone. *Avena fatua* grows wild in the valley. Therefore, it appears that oats were probably locally adopted by the ancient inhabitants as a fodder crop.

5.2. MILLETS

5.2.1. *Panicum miliaceum* (Figs. 5.14 and 5.15)

Caryopses are oval with a smooth, shining surface, 2.9 mm to 3.1 mm long and 1.5 mm to 2 mm broad, dorsal side slightly convex. A SEM of the surface shows that it is made up of smooth-walled parenchymatous cells. Caryopses were recovered from Indo-Greek phase at Semthan.

COMMENTS

The genetics of *Panicum* is not well understood. The common millet be-

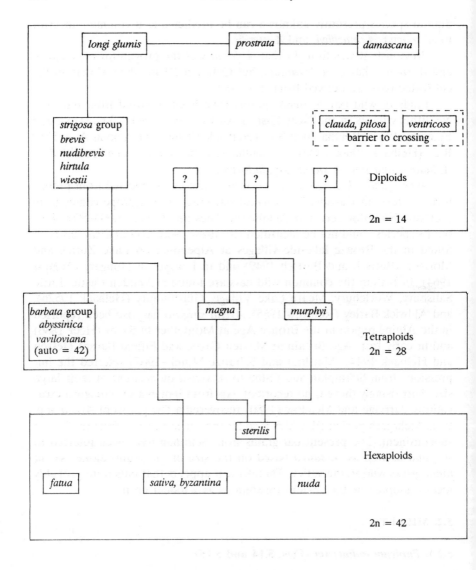

Text fig. 5: Species relationships in *fuana* spp.

Archaeological Evidence 131

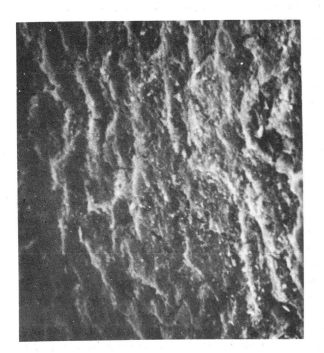

Figure 5.15: SEM of *Panicum*

Figure 5.14: *Panicum* sp. from Semtham.

longs to *P. miliaceum* and is not known in the wild state. On the basis of close morphological similarity, its progenitor is thought to be wild Abyssinian species *P. callosum* Hochst (Helbaek 1952c). *Panicum spontaneum* Lyssovex Zukovsking, which occurs as a weed in the crop in Central Asia, is probably a derivative rather than its progenitor (Purseglove 1974).

Panicum miliaceum originated in northern China and has been cultivated since earliest Neolithic times (Bishop 1933, Li 1970, Zhimin 1986). The earliest archaeological finds so far come from Neolithic of China, and Central and eastern Europe. It has been reported in the Aggtelek cave in Hungary (Bertsch and Bertsch 1949), in the Eisenburg settlement in Danubian I culture in Thuringia (Natho and Rothmaler 1957), at several Tripolye sites in Roumania (Gimbutas 1956), in the Neolithic Swiss lakeside villages (Bertsch and Bertsch 1949, in Neolithic contexts in Poland (Schultze-Motel 1968), and in Bronze Age sites in Italy (Helbaek 1956), Holland (Helbaek 1961), and Denmark (Helbaek 1952c). It has been found in Neolithic Yang Shao contexts in China (Watson 1969). In the Near East the earliest record is from Jemdt Nasr in Mesopotamia dating c. 3000 B.C. (Helbaek 1959b), followed by seventh century B.C. deposits at Nirmud (Helbaek 1966d).

So far there is no record of *Panicum* from prehistoric contexts in India. Recovery of only few grains from Semthan, though not a positive evidence of its cultivation, yet indicates that because of its earlier harvesting and drought endurance properties *Panicum* might have been adopted by the ancient inhabitants and was probably introduced from China, its centre of origin. The historical data confirm that some areas in the valley have been used for its cultivation during famine and drought conditions (Lawrence 1967).

5.2.2. Setaria spp. (figs. 5.16 and 5.17)

Caryopses are oval, longer than broad, glumed with tubercles on the surface, 2 mm to 2.5 mm long, and 1.2 mm to 1.5 mm broad. The SEM of the surface shows that it is made of sinous-walled cells. Grains were recovered from the Kushan phase at Semthan.

COMMENTS

The cultivated species *Setaria italica* L. is not known in the wild state and is believed to have been derived from *S. viridis* (L.) Beauv, a common weed in the Old World (Purseglove 1974). *Setaria italica* is of very early appearance in northern China and it is quite possible that its cultivation began in China (Li 1970, Zhimin 1986). It has been recorded from more than 20 sites in the Yellow River Valley, China (Zhimin 1986).

Setaria italica has been identified from Montellier and Buchs lakeside

Archaeological Evidence 133

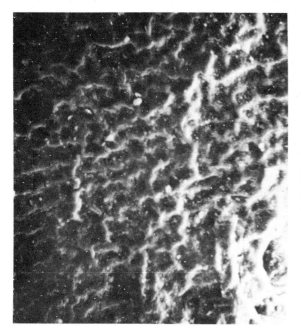

Figure 5.17: SEM of *Setaria* sp.

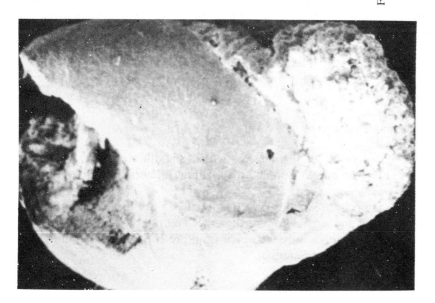

Figure 5.16: Broken *Setaria* sp. from Semthan.

settlements (Lee 1866), from several sites in Switzerland and from Hallstall in Austria (Neuweiler 1905). In the New World, *Setaria* dated 6000 to 5500 years B.P. is known from Ocampo Caves, Mexico, and Tehuacan Valley (Reed 1976).

Setaria viridis has been recorded from Neolithic deposits in the Aggtelek cave in Hungary, at Schussenthal in the Bronze Age settlement at Alpenquai on Lake Zurich (Bertsch and Bertsch 1949), and in the stomach of Grauballe man (Helbaek 1958a).

In India, Vishnu-Mittre and Savithri (1978) reported for the first time *Setaria* in the ancient plant economy from Surkotada, Gujarat belonging to *S. italica*, *S. viridis* or *S. verticillata* dating back to 1600 B.C. Wagner (1983) reported one carbonized and 100 uncarbonized *Setaria* grains from Oriya Timbo, a late Harappan site in Gujarat. However, there is no earlier record of *Setaria* spp. from prehistoric contexts of Kashmir. *Setaria viridis* and *S. glauca* are the two species of the genus which grow wild in Kashmir. Therefore, *Setaria* spp. might have been locally adopted by the inhabitants.

5.3. PULSES

5.3.1. Phaseolus mungo (fig. 5.18)

Seeds are oblong, covered with a thin, smooth seed coat, 4.0 mm to 4.8 mm long, 2.5 mm to 3.0 mm broad, with hilum lateral in position surrounded by a raised border which partially covers it.

Palisade cells are rather short, 58 μm to 67 μm in height, bulbous toward the inner end and narrow toward the outer end.

Seeds were recovered from N.B.P., Indo-Greek and Kushan phases at Semthan.

5.3.2. Phaseolus aureus (fig. 5.19)

Seeds are ovoid to oblong, covered with a smooth seed coat, 3.2 mm to 3.7 mm long, 2.5 mm to 2.6 mm broad. Hilum is lateral in position, oblong to oval in shape.

Palisade cells are rather short, 42 μm to 75 μm in height, bulbous at the inner end and narrow toward the outer end.

Seeds of *P. aureus* were recovered from Pre-N.B.P., N.B.P., Indo-Greek and Kushan phases at Semthan.

5.3.3. Phaseolus aconitifolius (fig. 5.20)

Seeds are very much compressed, oblong to oval, 3.3 mm to 5.1 mm long, 2.5 mm to 2.9 mm broad. Seed surface is smooth. Hilum is lateral in position. Palisade cells are short, 45 μm to 82.5 μm in height, bulbous at the inner end and narrow toward the outer end. The bulbous part of the cells has somewhat dense contents.
Seeds were recovered from Kushan phase at Semthan.

COMMENTS

Phaseolus aureus is of ancient cultivation in India but the plant is not found in the wild state. It is probably derived from *P. radiatus* L., which occurs wild throughout India and Burma (Purseglove 1977). Vavilov (1949–50) has proposed Indian and Central Asiatic centre of origin for the crop. The archaeological records are from Navdatoli Maheshwar 1550–1440 B.C. (Vishnu-Mittre 1962, 1974a), Diamabad (Kajale 1977b), and Apegaon (Kajale 1979).

Similarly, *P. mungo* is of ancient cultivation and has probably originated in India but is not known in the wild state. It probably originated from *P. trinervius* Heyne or *P. sublobatus* Roxb., which occur wild in India (Vavilov 1949–50, Purseglove 1977). Hitherto the earliest record is from Banawali c. 2300 B.C. (Lone et al. 1987), followed by Navdatoli Maheshwar 1550–1440 B.C. (Vishnu-Mittre 1962, 1968a, 1974a), Diamabad (Kajale 1977b), Atranjikhera (Chowdhury et al. 1977), Apegaon (Kajale 1979), Nevasa and Inamgaon (Kajale 1977a). There is no earlier record from Kashmir.

So far as *P. aconitifolius* is concerned it originated in the Indian centre of origin of Vavilov (1949–50). It is a native of India, Pakistan, and Burma, where it grows wild (Purseglove 1977). A perusal of literature does not reveal any record of its presence in the archaeological contexts of India or elsewhere.

The evidence presented above indicates that various species of *Phaseolus* have been introduced into the valley from the Indian centre of origin of Vavilov (1949–50), at various stages of cultural development.

5.3.4. *Lens culinaris* (fig. 5.21)

Seeds are flat and circular, 3 mm to 4 mm in diameter, 1.6 mm to 2.0 mm in thickness. Seed surface is smooth. Hilum is lateral, oval, in line with the seed surface, and 1 mm in size.
Palisade cells are short, 25 μm to 34 μm in height, bulbous at the inner end and narrow toward the cuticular end.

136 Palaeoethnobotany

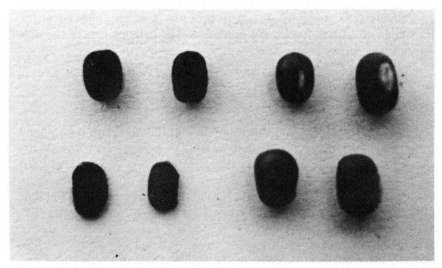

Figure 5.18: Archaeological *Phaseolus mungo* seeds (left) from Semthan and extant seeds (right).

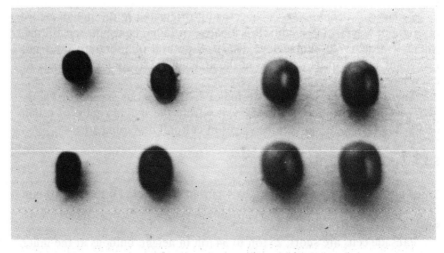

Figure 5.19: Archaeological (left) and extant (right) *Phaseolus aureus*.

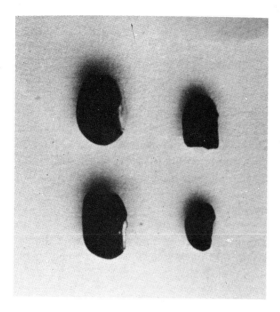

Figure 5.20: Ancient (right) and extant (left) *Phaseolus aconitifolius*.

Seeds were recovered from all the four phases at Burzahom and Pre-N.B.P, N.B.P, Kushan, and Hindu Rule phases at Semthan.

COMMENTS

Lentils are diploid with 2n = 14 (Hector 1936). The cultivated species *Lens esculenta* syn. *L. culinaris* was thought to have been derived from the wild *L. nigricans*, which is native to southeast Europe and western Asia (Bertsch and Bertsch 1949), but comparative morphology and observations on natural hybridization have revealed that *Lens orientalis* (Boiss) Hand. is the ancestor of the cultigen which is distributed mainly in Turkey, Syria, Israel, northern Iraq, and western and northern Iran (Zohary 1972, 1976).

Lentils were definitely associated with the start of agriculture in the Near East (Zohary 1972, 1973, 1976, Zohary and Hopf 1973, 1986, Renfrew 1973), which confirms the belief that the centre of origin of lentil is in the Near East (Vavilov 1949–50). Carbonized lentil seeds have been reported from Jericho c. 7000 B.C. (Hopf 1969); Jarmo c. 6750 B.C. (Helbaek 1960a); Tape Sabz 5500–5000 B.C. (Hole and Flannery 1967); Ali Kosh 7500–5600 B.C. (Helbaek 1966e); and Hacilar 5800–5600 B.C. (Helbaek 1966e). Significantly, some of the fifth millennium B.C. remains were larger than the wild form and attain 42 mm diameter, which is an obvious development under domestication (Zohary 1976). In Europe lentils occur at Argissa c. 6000–5000 B.C. (Hopf 1962); Ghediki c. 6000–5000 B.C. (Renfrew 1966); Tell Azmak c. 5000 B.C. (Renfrew 1969); and some other sites in Switzerland, the Mediterranean basin, and central Europe (Renfrew 1975, Zohary 1976).

In India lentil is recorded at Chirand 1800 B.C. (Vishnu-Mittre 1972a); Navdatoli Maheshwar 1550–1400 B.C. (Vishnu-Mittre 1962, 1968a, 1974a); Bokhardan 200 B.C. (Kajale 1974a); Diamabad 2200–1000 B.C. (Kajale 1977); and Ter 150 B.C.–100 A.D. (Vishnu-Mittre et al. 1971). Thus lentils are of early cultivation in West Asia and southern Europe and from these areas spread northward in Europe, eastward to India and southward to Ethiopia (Purseglove 1977). The present evidence from Kashmir clearly indicates that the lentil has been introduced from West Asia and has been utilized from the very dawn of agriculture here, along with wheat and barley.

5.3.5. *Pisum sativum* (fig. 5.22)

Seeds are large, round to ovoid, and 4.5 mm to 8.5 mm in diameter. Seed surface is thin and smooth. Hilum lies level with the seed surface and is oblong and 1.6 mm to 2.8 mm in size.

Palisade cells are rather long, 67.5–83 μm in height. The inner end of the cells is corrugated and broad while the outer end is narrow and smooth.

Seeds were recovered from Neolithic, Megalithic and Post-Megalithic phases at Burzahom and from N.B.P., Kushan, and Hindu Rule phases at Semthan.

COMMENTS

Neither the wild progenitor nor the early history of the pea crop is known (Davies 1976). All the peas are diploid with $2n = 14$. Zukovskj (1950) suggests that the wild ancestor may be *Pisum elatius* and *P. arvense* (*P. sativum* var. *arvense*) may be an intermediate form. Ben-ze'ev and Zohary (1973) and Davis (1970) maintain that the genus consists of only *P. fulvum* and *P. sativum* (*P. elatius*, *P. humile* and *P. sativum* being members of a single species).

Vavilov (1949–50) proposed central Asiatic and Near Eastern centre of origin for *P. sativum*. The earliest archaeological finds are those from Jericho c. 7000 B.C. (Hopf 1969), Jarmo c. 6750 B.C. (Helbaek 1960a), and Can Hasan c. 5250 B.C. (Renfrew 1968), all belonging to *P. sativum* var. *arvense* (Renfrew 1973). At aceramic Hacilar c. 7000 B.C. and at Catal Huyuk 5850–5600 B.C., *P. elatius* was predominant (Helbaek 1964a). In Europe peas occur at Ghediki c. 6000–5000 B.C. and Sesklo 6000–5000 B.C. and at some other sites (Renfrew 1966, 1973).

Archaeological Evidence 139

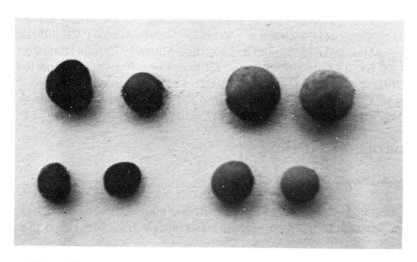

Figure 5.21: Archaeological (left) and extant (right) *Lens culinaris*.

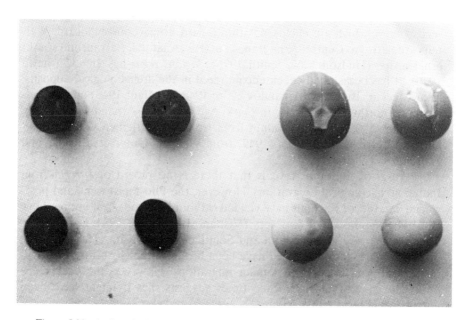

Figure 5.22: Archaeological (left) and extant (right) *Pisum sativum*.

The archaeological records of *P. arvense* from India are from Harappa 2250 B.C. (Vats 1941), Chirand c. 1800 B.C. (Vishnu-Mittre 1971, 1972a), Navdatoli Maheshwar 1550–1440 B.C. (Vishnu-Mittre 1962), and Diamabad 2200–1000 B.C. (Kajale 1977b). In the Kashmir valley *P. arvense* is reported from Gofkral c. 2100 B.C. (Sharma 1982). From the present evidence it appears that the crop was introduced into the valley from Central Asia or West Asia sometime during the third millennium B.C.

5.4. HORTICULTURAL FRUITS

5.4.1. *Juglans regia* (figs. 5.23 and 5.24)

Pieces of endocarp were found 7 mm to 20 mm long, with outer surface irregularly grooved, forming reticulations, and inner surface with raised projections.

Endocarp is made up of sclereids. The majority of sclereids are asterosclereids, others are brachysclereids, 62.5–102 μm in diameter.

Endocarps were recovered from Neolithic I, Neolithic II, and Megalithic periods at Burzahom and Pre-N.B.P., N.B.P., Indo-Greek, and Kushan phases at Semthan.

COMMENTS

Several species are cultivated or exploited from the wild for their edible seeds (nuts), which are enclosed within a hard drupaceous endocarp. Walnuts are native to Central Asia, Iran, Caucasus, Anatolia, Balkan, and southern Europe (Hudson 1962, Smith 1976). *Juglans regia* has 2n=32 (Smith 1976). It has occasionally been encountered in the archaeological deposits, as at Meckur, Bulgaria (Arnaudov 1936, 1948–49, Gaul 1948) and at Sadowetz in northwest Bulgaria, in Dark Age deposits (Arnaudov 1937–38). Neuweiler (1905) reported walnut from Iron Age settlements at Fontinellato and Bertsch and Bertsch (1949) from Wangen, Untersee, Bleiche, and Haltnau.

Perusal of literature reveals that there is no record of *J. regia* from Indian archaeological excavations. However, the Plio-Pleistocene and post-glacial deposits of Kashmir have yielded carbonized woods, leaf impressions, and pollen grains of *Juglans* spp. (Vishnu-Mittre 1965, 1984, Lone 1987). Thus its recovery at Burzahom and Semthan is significant. The foregoing evidence suggests that it could have been brought from its centre of origin in Central Asia along with human migrations. However, the occurrence of *Juglans* since Plio-Pleistocene times suggests that it might have been domesticated locally in Kashmir.

Archaeological Evidence 141

Figure 5.23: Endocarps of *Juglans regia*.

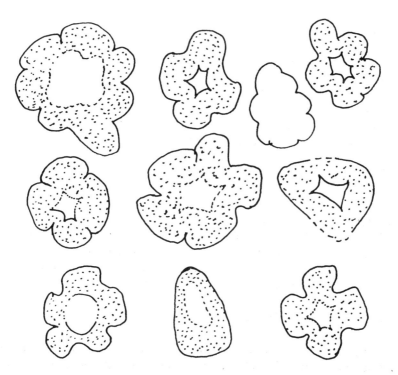

Figure 5.24: Sclereids from ancient *Juglans regia* 400x.

5.4.2. *Prunus persica* (fig. 5.25)

Pieces of endocarp were found, 4 mm to 18 mm long, thick and hard, with deep ridges on the outer surface and a smooth inner surface.

Sclereids were of two types: macrosclereids 125 μm to 187 μm long and 8 μm to 65 μm broad, and brachysclereids 48 μm to 85 μm in diameter.

Endocarps were recorded from Neolithic I, Megalithic and Post-Megalithic periods at Burzahom and Indo-Greek, Kushan, and Hindu Rule phases at Semthan.

5.4.3. *Prunus armeniaca* (fig. 5.26)

Endocarp pieces were found, 2 mm to 19 mm long, comparatively thin, with characteristic ripple markings on the outer surface and a smooth inner surface.

Sclereids were of two types: macrosclereids 112 μm to 165 μm long and 25 μm to 65 μm broad, and brachysclereids 50 μm to 62 μm in diameter.

Recorded from Neolithic I, Neolithic II, and Post-Megalithic periods at Burzahom and all the five phases at Semthan.

5.4.4. *Prunus amygdalus* (fig. 5.27)

Endocarp pieces were found, 5 mm to 7 mm long, with outer surface deeply pitted, pits in rows, and a smooth inner surface.

Sclereids are of two types: macrosclereids, 93 μm to 122 μm long and 20 μm to 61 μm broad, and brachysclereids 50 μm to 62 μm in diameter.

Endocarps were recorded at Neolithic II and Post-Megalithic phases at Burzahom. They were not found at Semthan.

5.4.5. *Prunus domestica* (fig. 5.28)

Endocarps are oval to globular, thin, 10 mm to 10.5 mm in diameter, with a smooth outer as well as inner surface.

Macrosclereids are 125 μm to 287 μm in length and 20 μm to 45 μm in breadth; brachysclereids are 35 μm to 47 μm in diameter.

Endocarps are recorded from Megalithic and Post-Megalithic periods at Burzahom. They were not recorded at Semthan.

5.4.6. *Prunus cerasus* (fig. 5.29)

Endocarp pieces were found, 2 mm to 10 mm long, characterized by a smooth outer surface devoid of ripples, grooves or pits and a smooth inner surface.

Archaeological Evidence

Figure 5.25: Archaeological *Prunus persica* endocarps.

Figure 5.26: Ancient *Prunus armeniaca* endocarps.

Figure 5.27: Ancient *Prunus amygdalus* endocarp.

Macrosclereids are 102 μm to 162 μm long and 17 μm to 55 μm broad. Some sclereids have pitted walls.

Endocarps were recorded from N.B.P. phase at Semthan. They were not recorded at Burzahom.

COMMENTS

It seems probable that the first diploid *Prunus* species arose in Central Asia (Vavilov 1949–50, Watkins 1976).

Watkins (1976) believes that *P. armeniaca*, the apricot, has its primary centre of origin in Western China and its secondary centre of origin in Western Asia. However, there is no archaeological record from India or elsewhere except Kashmir, where it has been reported from Gofkral c. 2000 B.C. contexts (Sharma 1982). In the light of these facts it appears that this fruit has been introduced into the valley from central Asia or Western China.

Prunus persica, the peach, is also believed to have originated in Western China on botanical grounds (Vavilov 1949–50, Li 1970, Watkins 1976). However, there is no archaeological record. Its recovery from the Kashmir archaeological excavations and its continued cultivation nowadays makes it appear to have been introduced from China.

As far as *P. cerasus* is concerned its place of domestication is not known, but it is believed to have evolved from *P. fruticosa* in western and central Asia (Watkins 1976). So far it has not been found in the archaeological deposits. Interestingly *Prunus* spp. are represented in the Pleistocene deposits of the valley (Vishnu-Mittre 1984, Lone 1987). Further, *P. cerasus* does grow wild in the valley at present. Therefore, it might have been collected locally by ancient man.

Almond is closely related to peach but became established in a separate centre of origin in an area extending from central to Western Asia (Vavilov 1949–50, Watkins 1976). Therefore, it must have been brought from there along with the migration of people.

Prunus domestica has its centre of origin in Europe (Vavilov 1949–50, Li 1970, Watkins 1976). However, it grows wild in Kashmir and therefore must have been available locally.

5.4.7. *Vitis vinifera* (Wild.)

Seeds are pyriform with a short stalk, and an oval-circular chalazal scar or 'shield' on the dorsal surface. On the ventral side two narrow, deep furrows flank a central longitudinal ridge.

Seeds were recorded from Neolithic II phase at Burzahom only.

Archaeological Evidence 145

Figure 5.28: Ancient *Prunus domestica* endocarps.

Figure 5.29: Ancient *Prunus cerasus* endocarps.

COMMENTS

The estimated 10,000 cultivars of the Old World are thought to be derived from the single wild species *Vitis vinifera* of Middle Asia, still found from northeastern Afghanistan to the southern borders of the Black and Caspian Seas (Olmo 1976). Cultivation of wine grape was under way in the Near East as early as the fourth millennium B.C. The products of vine were exported westward from very early times to be followed later by the practices peculiar to viticulture and by domesticated varieties (Helbaek 1959b). It is from the Near East that *Vitis* is believed to have reached India and other parts of the world (Olmo 1976).

The grapevine, *V. vinifera*, is mentioned in the Bible and all ancient Hebrew writings as 'gafen', originating apparently from a verbal radical 'Kafan' meaning to bend and curl, in reference to vine stems and tendrils. The grapevine has been cultivated in the Holy Land for many centuries. Vines in the wild state have been found in Europe, Greece, Anatolia, Iran, and northern India. The authorized version of the Bible translating Isaiah 5:2 renders 'beoushim' as wild grapes; most Hebrew scholars, however, are of the opinion that it means unripe grapes or grapes that will never ripen, but in no case does it mean wild grapes.

In France and Italy, fossil vines were discovered as far back as the beginning of the Quaternary Age. Leaves and seeds indicate the existence of another species, *V. silvestris*, during the Tertiary Age in Switzerland, Italy, Britain, and Iceland. Seeds found in the middens of lake dwellings of southern central Europe prove that the grape has been a human food from very early times.

It has been assumed that the origin of the European vine and of its culture should be placed in the Middle East at the northern tip of Southwest Asia in the vicinity of the Caspian Sea. The Bible refers to a vineyard planted by Noah in the same region on the mountains of Aravat, 'And Noah began to be a husbandman and he planted a vineyard and he drank the wine' (Genesis 9:20–21). The credit of the origin of *V. vinifera* is presently given to an area extending from Southwest Europe to Western India. Archaeological records are from the Bronze Age levels at Jericho and Lancist. Seeds from the Iron Age have been found in many excavations in Isreal.

Grapes have been reported from Mehrgarh (Bronze Age) and Nausharo (Indus levels) by Costantini (1984). Charcoals of *Vitis* have been identified from Mehrgarh and Nausharo (Thiebault 1986). They have also been identified from Tape Yahya, Iran (Lamberg-Karlovsky and Beale 1986).

In view of the above discussion one is led to believe that origin of the grapevine of Kashmir lies in the Middle East and that it might have been introduced via Central Asia.

5.5. WEED SEEDS

5.5.1. *Lithospermum arvense* (fig. 5.30)

Seeds are ovoid-truncate with a ridge on one side and a broad flattish scar at the base, and are 2 mm to 3 mm long, and 1.5 mm to 2 mm broad. The surface is crusty, hard, papillate, and roughened with tubercles. Papillae as revealed by SEM are 70 μm to 150 μm long and 37 μm to 50 μm broad. Seeds were recorded from Neolithic II, Megalithic, and Post-Megalithic periods at Burzahom and N.B.P., Indo-Greek, Kushan, and Hindu Rule phases at Semthan.

5.5.2. *Galium tricorne* (figs. 5.31 and 5.32)

Seeds are globose to reniform, hollow-centred, 1.5 mm to 2.5 mm in diameter. The seed surface is smooth. Seeds were recorded from Neolithic II phase at Burzahom and Kushan and Hindu Rule phases at Semthan.

5.5.3. *Galium asperuloides* (fig. 5.33)

Seeds are globose to reniform, hollow-centred, and 1.7 mm to 2.6 mm in diameter. Seed surface is bristled, and papillae are scarce. Seeds were recorded from Kushan phase at Semthan.

5.5.4. *Galium aparine* (fig. 5.34)

Seeds are globose to reniform, hollow-centred, 1.4 mm to 2.5 mm in diameter. Seed surface is bristly with dense papillae. Seeds were recorded from Neolithic I, Megalithic, and Post-Megalithic phases at Burzahom and Kushan phase at Semthan.

5.5.5. *Vicia/Lathyrus* spp. (fig. 5.35)

Seeds are small, globose to rounded, and 2 mm to 3 mm in diameter. Position of hilum is clear on the seed surface, and the surface is smooth. Seeds were recorded at Neolithic I, Megalithic, and Post-Megalithic phases at Burzahom and N.B.P., Indo-Greek, Kushan, and Hindu Rule phases at Semthan.

5.5.6. *Medicago* spp. (fig. 5.36)

Seeds are half-moon-shaped with rounded ends, compressed falcate, and 2.5 mm to 3 mm long. Position of hilum is clear on the lateral side. Seeds were

Figure 5.30: Archaeological seeds of *Lithospermum arvense*.

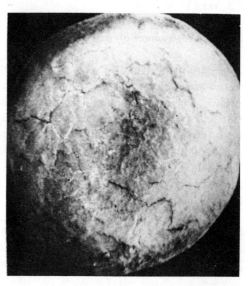

Figure 5.32: SEM of seeds of *Galium* tricorne.

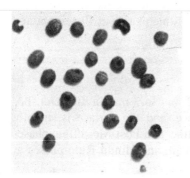

Figure 5.31: Archaeological *Galium* spp. seeds.

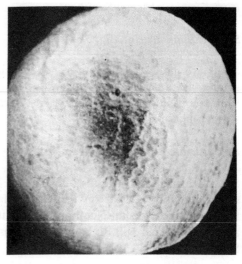

Figure 5.33: SEM of seeds of *G. asperuloides*.

recorded from Neolithic I and Post-Megalithic phases at Burzahom and Kushan phase at Semthan. One seed from Burzahom was uncarbonized.

5.5.7. *Melilotus albus*

Seeds are compressed, ovoid, truncate at the broader end, 2 mm long, and 1 mm broad, with small circular hilum on the edge near one end. Seeds were recorded from Neolithic II and Megalithic phases at Burzahom and Hindu Rule phase at Semthan.

5.5.8. *Ipomoea* spp. (fig. 5.37)

Uncarbonized seeds were found, yellowish-brown, thick wedge-shaped, with the two inner faces equal and outer face broad and rounded. Seeds are 3.5 mm to 5 mm long and 3 mm broad. Seed surface is smooth, with horseshoe-shaped scar fully exposed on the ventral side. Seeds were recorded from Neolithic I, Neolithic II, and Post-Megalithic phases at Burzahom. They were not found at Semthan.

5.5.9. *Astragalus* spp. (fig. 5.38)

Seeds are reniform, 3 mm to 3.5 mm long and 2.1 mm to 2.9 mm broad, with a smooth surface and lateral hilum. Seeds were recorded from Neolithic I, Megalithic, and Post-Megalithic phases at Burzahom. They were recorded at Semthan.

COMMENTS

The weed seeds are of great significance to agriculture and have played an important role in plant domestication. The evolution of weeds often parallels that of crops. Both weeds and crops often begin with a common progenitor and many cultivated plants have one or more companion weed races (Langenheim and Thimann 1982).

Any disturbance of natural vegetation provides growth opportunities for plants not normally dominant. In agricultural clearing selective weeding further favours particular small or low-growing plants. This phenomenon has been much studied and complexes of weeds in fields of different crops in most parts of the world are well known. Their occurrence in excavated pollen spectra, particularly in macroscopic collections, sometimes indicates not only cultivation but even the kind of crops being grown (Alexander 1969). Weed seeds have been associated with prehistoric deposits all over the world. Renfrew (1973) has provided an exhaustive catalogue of the wild plants recovered from archaeological excavations of the Near East and

150 Palaeoethnobotany

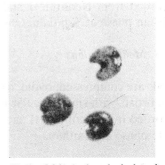

Figure 5.36: Archaeological seeds of *Medicago* spp.

Figure 5.34: SEM of seeds of *G. aparine*

Figure 5.37: Archaeological seeds of *Ipomoea* spp.

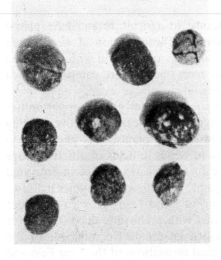

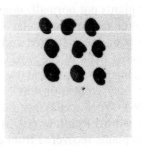

Figure 5.35: Archaeological seeds of *Vicia/Lathyrus* spp.

Figure 5.38: Archaeological *Astragalus* spp.

Europe. Weed seeds have also been identified from some archaeological excavations in India (Vishnu-Mittre, 1962, 1966, 1968a, Sharma 1982). The seeds recovered from Burzahom and Semthan belong to *Lithospermum arvense, Galium tricorne, G. asperuloides, G. aparine, Ipomoea* spp., *Astragalus* spp., *Medicago* spp., *Melilotus albus*, and *Vicia/Lathyrus* spp. Either the association of these weed seeds with cereals and other food plants is deliberate, indicating a primitive stage of farming when both wild and cultivated plants were harvested, or the weed seeds were merely contaminants with the harvested crops. The third possibility is that these weeds were growing in or around the vicinity of the site and somehow found their way into the floor. All these genera are among the common weeds of cultivated lands of the valley today.

5.6. WOODS

5.6.1. *Platanus orientalis* (figs. 5.39 and 5.40)

The charcoals that could be ascribed to the taxon *Platanus orientalis* came from the Megalithic period at Burzahom and Early Historic period at Semthan.

The wood is semi-diffuse to diffuse porous. Growth rings are distinct, marked by a band of firm texture with a few pores on the outer edge of each ring, under a lens. Vessels are small, frequently crowded or clustered, and occasionally solitary. They are oval to round and 30 μm to 80 μm in diameter. Inter-vessel pits are small, alternate and crowded. Tyloses are occasionally present. Parenchyma is paratracheal and metatracheal diffuse. Paratracheal parenchyma is scanty. Fibres are thin-walled to moderately thick-walled, angular to oval, and 8 μm to 15 μm in diameter. Rays are broad, equidistant, multiseriate, essentially homogeneous, 200 μm to 450 μm and 12 to 23 cells in height and 47 μm to 103 μm and 3 to 9 cells in width. Between the broad cells are occasional fine ones.

The above anatomical characters and further comparison with the extant wood confirm that the charcoals belong to *P. orientalis*.

COMMENTS

The oriental plane (*Platanus orientalis* L.) is a large, graceful, deciduous tree cultivated in Kashmir and the northwest Himalayas. The tree highly valued as an ornamental plant in Kashmir, is found in the Balkan peninsula southward from 42°N, Kriti; Often planted elsewhere (Tutin et al. 1964), it is also reported to be a native of the Mediterranean region (Anonymous 1969a, Brandis 1971). In Kashmir the tree is known as 'Chinar' or 'Buin' and is very closely associated with the culture and folklore. Its introduction usu-

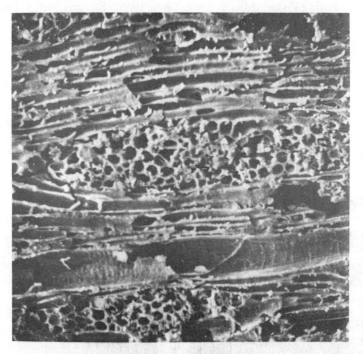

Figure 5.40: T.L.S. *Platanus orientalis* 200x.

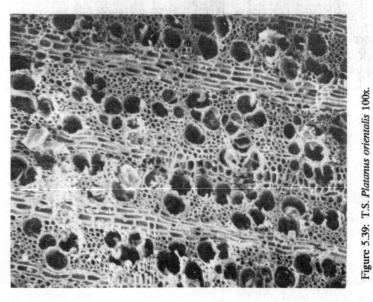

Figure 5.39: T.S. *Platanus orientalis* 100x.
*Images in figs. 5.39–5.86 are taken under scanning electron microscope.

Archaeological Evidence 153

ally has so far been ascribed to the Moghul emperors Jehangir (1605–1627 A.D.) and Shah Jahan (1627–1658 A.D.), who brought it from Central Asia (Lawrence 1967).

But the present evidence from Burzahom (1000–600 B.C.) and Semthan (500–1000 A.D.) has revealed that this tree has been in existence here centuries before the advent of the Moghuls. The charcoals bear ample testimony to its much earlier introduction into the Kashmir valley. Some references in this regard could also be found in the literature. Lal Ded (1320–1390 A.D.), the famous mystic poetess of Kashmir, in an epigram compares 'virtuous and loving wife to the cool and refreshing shade, on a hot summer day, of a Buin'. Lal Ded lived about two centuries before the advent of the Moghuls in Kashmir.

Further, the *Akbarnama*, written by Abul Fazl, a scholar in the regime of the Moghul emperor Akbar the Great, says that 'the emperor took 34 persons inside the hollow trunk of an aged Chinar'. Similarly, Jehangir in his memoirs has mentioned a huge Chinar tree in the hollow of which he and seven of his companions could be comfortably accommodated. Evidently such grand trees must have been hundreds of years old. Now the archaeological evidence from Burzahom and Semthan has virtually proved that the plane tree has been growing in Kashmir since much earlier times.

How and when the tree reached India still remains obscure. But there may be two possibilities: either its introduction is an example of extension of the Mediterranean flora into the valley, or the cultural contacts of the prehistoric people might be responsible. Whatever might have happened in the past, the introduction of Chinar in the Kashmir valley, centuries before the advent of the Moghuls, has opened new vistas in the early history of the introduction of exotic species. Nevertheless, the introduction of Chinar has so far not been explored in the light of botanical research.

5.6.2. *Morus alba* (figs. 5.41 and 5.42)

The evidence of *Morus alba* has come from the Early Historic period at Semthan.

The wood is ring porous. Growth rings are distinct. Vessels in the spring wood are larger than those in the summer wood; oval to rounded; solitary, in pairs, or in groups of up to five; and thin-walled. Larger vessels are 50 μm to 90 μm in diameter and smaller ones 20 μm to 40 μm. Some are occluded with tyloses. Inter-vessel pits are small and oval to orbicular. Parenchyma is paratracheal and metatracheal diffuse. The paratracheal parenchyma surrounds the vessels and forms continuous bands, i.e., paratracheal confluent. Fibres are thin walled to moderately thick-walled, angular, oval to round, and 10 μm to 15 μm in diameter. Rays are fairly wide, normally spaced, 1-6 seriate, homogeneous to heterogeneous.

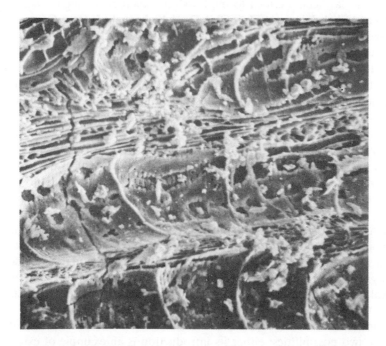

Figure 5.42: T.L.S. *Morus alba* 200x.

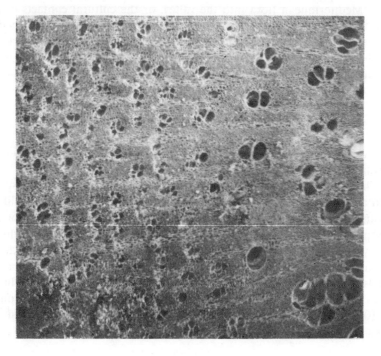

Figure 5.41: T.S. *Morus alba* 10x.

COMMENTS

The present evidence is the first and only report of mulberry tree (*Morus alba* in the archaeological excavations of Kashmir. *M. alba* is of very ancient cultivation in China and is not indigenous to India. It is commonly cultivated in Baluchistan, Afghanistan, the northern part of the trans-Indus territory, the Punjab plains, Kashmir, the northwest Himalayas, Europe, Central Asia, and China (Brandis 1971). Apparently the tree has been introduced into Kashmir from China mainly for silkworm rearing (sericulture).

The genesis of sericulture in Kashmir is as yet an unsolved historical riddle. Many theories have been put forth. Some assert that sericulture has an indigenous origin in Kashmir (Ganju 1945), whereas others believe that it is a fifteenth to sixteenth century introduction in Kashmir (Mirza Haider 1973). The first notion is based on the reverence paid to mulberry trees by the Hindus at two religious ceremonies, namely the 'Bhairwa Pooja' and the 'Yagneopavita ceremony' since very ancient times. This has led Ganju (1945) to conclude that silkworm rearing based on mulberry leaves was carried out by the people of Kashmir since very ancient times.

The second theory is based on the fact that mulberry trees were growing in Kashmir in abundance during the sixteenth century. To quote Mirza Haider, 'Among the wonders of Kashmir are the quantities of mulberry trees cultivated for their leaves from which silk is obtained.'

In the light of the present discovery of *M. alba* in the contexts dating 500–1000 A.D. and lack of any record of the genus in Plio-Pleistocene and post-glacial deposits of the valley, it is clear that sericulture is neither indigenous nor a fifteenth or sixteenth century introduction. It being an established fact that sericulture first originated in China (Encyclopedia Britannica Vol. 20 p. 519, Radziniski 1979), the mulberry must have been transported into Kashmir from there. It is very difficult to give an exact date of its diffusion, at this stage at least. However, the present evidence has led us to 500–1000 A.D. The silk route could have been followed for the purpose (see fig. 6.1). During this period Kashmir had developed very close contacts with Central Asia and China (Ahmad 1986, Shali 1986).

5.6.3. *Quercus* spp. (figs. 5.43–5.45)

Although the exact species of *Quercus* could not be determined, the variability clearly indicates two distinct species. These are regarded as *Quercus* sp. I and *Quercus* sp. II. *Quercus* sp. I has been revealed from the Pre-N.B.P. period at Semthan and Neolithic I and Neolithic II periods at Burzahom. *Quercus* sp. II came from the N.B.P. period at Semthan.

QUERCUS SP. I.

The wood is ring porous. Vessels are medium-size to large, in pairs or in multiples of up to four cells. The early wood vessels form more or less continuous rows in the form of radial strings. Vessels are 75 μm to 150 μm in tangential diameter. Inter-vessel pits are moderately large and oval to orbicular. Parenchyma is sparse, metatracheal and diffuse. Fibres are thin-walled to moderately thick-walled, oval to angular, and 3 μm to 10 μm in diameter. Rays are unstoried, 2–5 seriate, homogeneous to heterogeneous.

QUERCUS SP. II

The wood is ring porous. Vessels are very large in the spring wood, 90 μm to 180 μm in tangential diameter; summer wood vessels are smaller, 40 μm to 60 μm in diameter. Vessels are rounded, solitary, visible to the naked eye, and thin-walled. Inter-vessel pits are orbicular to elliptical, 3 μm to 6 μm in diameter, and vessel segments are annular. Parenchyma is paratracheal as well as metatracheal diffuse. Paratracheal parenchyma is sparse and restricted to a few cells and does not form a sheath. Fibres are thick-walled, angular to rounded, and 10 μm to 15 um in diameter. Rays are of two types: uniseriate rays and broad aggregate rays of oak type. Broad rays are 8–10 seriate and 75 μm to 100 μm in width. Rays are unstoried and homogeneous.

COMMENTS

Oaks (*Quercus* spp.) present the most vexing problem among the floral elements of the valley. They are almost absent in the present-day flora of the valley (Royle 1839, Drew 1875, Parker 1924, Lambert 1933, de Terra and Paterson 1939, Puri 1948, Rao 1960). However, the megafossil evidence has revealed their presence during Pliocene and Pleistocene times in the valley (de Terra and Paterson 1939, Puri 1948, Vishnu-Mittre 1965, Lone 1987). The palynological studies have confirmed these observations (Vishnu-Mittre et al. 1962, Gupta et al. 1985, Sharma et al. 1985, Sharma and Gupta 1985). Of major interest is the evidence of oaks in the pollen analysis of post-glacial deposits in the valley (Singh 1963, Vishnu-Mittre and Sharma 1966, Sharma and Vishnu-Mittre 1968, Singh and Agrawal 1976, Vishnu-Mittre 1984) and in some bogs (Dodia et al. 1985).

Thus the identification of *Quercus* sp. in the archaeclogical woods is quite significant. Though the exact species could not be identified, the woods do not belong to the exotic European Oak *Quercus robur* L. that has been planted in the valley in recent years (Vishnu-Mittre 1963). Hence the charcoals belong to some indigenous species. Some stray patches of

Archaeological Evidence 157

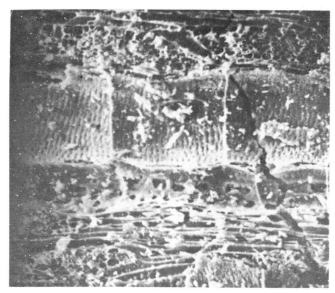

Figure 5.44: T.S. *Quercus* sp.I. 200x.

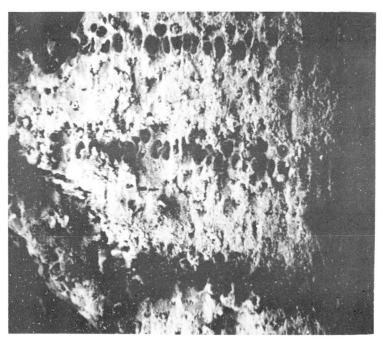

Figure 5.43: T.S. *Quercus* sp. 50x.

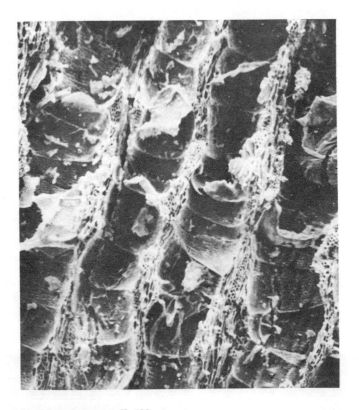

Figure 5.45: T.L.S. *Quercus* sp.II. 100x.

Q dilatata and *Q semicarpifolia* do occur in the present-day valley forests (Vishnu-Mittre 1963, personal observation).

The dominance of oaks in the valley during Pliocene and Pleistocene times and their near absence today has largely been explained by the uplift of the mountain barrier: the Pir Panjal, which flanks the valley to the south and southwest. This has acted as an effective barrier and prevented the monsoon from entering the valley, resulting in extremely low precipitation (Puri 1957, Puri et al. 1983). However, the presence of oaks in the post-glacial deposits and now in the archaeological deposits does not support this thesis as there is no evidence to suggest that the mountain barrier subsided substantially during the post-glacial period, thus allowing the monsoons to enter the valley and create ideal climatic conditions for the immigration of oaks. Further, the climatic requirements of some species of oaks distributed in the western Himalayas show a very great range and reduction in precipitation could not have been detrimental to the extent of their extinction. Some other factors must also be involved.

The present-day distribution of the various species of oaks is as follows (Troup 1921, Pearson and Brown 1931, Brandis 1971):

Quercus semicarpifolia grows in Kuram valley from 9,000 ft to 11,000 ft, in the Himalayas from 8000 ft to 10,000 ft, and in eastern Manipur, on the Burma frontier. *Quercus incana* grows in the north-western Himalayas, and eastward as far as Nepal at 4,000 ft to 8,000 ft. *Quercus glauca* grows in the valleys of the outer Himalayas ascending up to 6,000 ft. *Quercus dilatata* grows in the Kuram valley at 7,000 ft to 8,500 ft. and in the northwestern Himalayas at 5,000 ft to 9,000 ft.

In view of these facts it appears that some patches of indigenous oaks must have been growing near the sites or at accessible places in the forests wherefrom the inhabitants collected the wood. This further provides a clue that in addition to climatic changes some other factors like biotic factor might also have been involved in the reduction of oaks from the valley.

5.6.4. *Betula utilis* (figs. 5.46 and 5.47)

Charcoals belonging to birch (*Betula utilis*) were recovered from Neolithic I period at Burzahom and N.B.P. and Kushan periods at Semthan.

The wood is diffuse porous. Growth rings are distinct. Vessels are almost of uniform size, evenly distributed in the grain, usually solitary, rarely in pairs, thin-walled to medium thick-walled, 30 μm to 110 μm in tangential diameter. Inter-vessel pits are medium-size and oval to orbicular. Parenchyma is metatracheal diffuse and paratracheal and scanty, paratracheal parenchyma not forming a sheath. Fibres are thin-walled to moderately thick-walled, angular, and 10 μm to 25 μm in diameter. Rays are normally spaced, unstoried, and of two types—uniseriate and 2-6 seriate; they are essentially homogeneous, up to 22 cells and 250 μm in height and 75 μm in width.

COMMENTS

The present-day distribution of *Betula utilis* in the Indian region is in the Kuram valley from 10,000 ft to 11,000 ft and in the Himalayas from 10,000 ft to 14,000 ft. Birch today occurs only at high altitude in the Kashmir valley. The retrieval of its charcoals at Burzahom and Semthan suggests that the tree might have been growing at lower elevations in the past. The names of archaeological sites such as Burzahom ('Burz' means birch in Kashmir) also suggest the occurrence of birch woods in the valley proper in the recent historical past. Its recent extermination from the valley proper must obviously be due to human influence. Its destruction may be attributed to the removal of its leaves to feed sheep and goats and removal of its bark to provide the famous 'bhojpatra' for writing manuscripts.

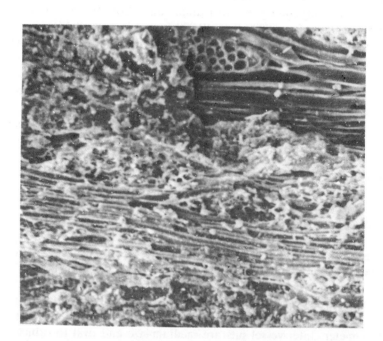

Figure 5.47: T.L.S. *Betula utilis* 200x.

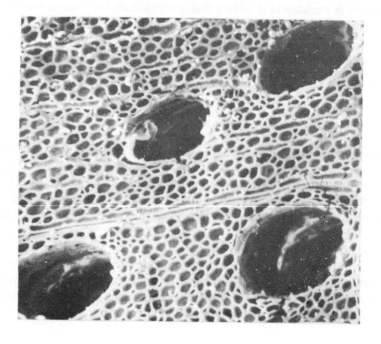

Figure 5.46: T.S. *Betula utilis* 200x.

5.6.5. *Juglans* spp. (figs. 5.48 and 5.49)

Walnut (*Juglans* spp.) charcoals have been recovered from Burzahom (Neolithic I and Post-Megalithic periods) and Semthan (Pre-N.B.P. and Kushan periods).

The wood is semi-ring to diffuse porous. Growth rings are not distinct. Vessels are ovoid to spherical, solitary or in multiples of 2 or 3, and 35 μm to 100 μm in tangential diameter. Tyloses are occasionally present. Parenchyma is apotracheal diffuse and paratracheal, and paratracheal parenchyma is scanty. Fibres are thin-walled to medium thick-walled, oval to angular, and 10 μm to 20 μm in diameter. Rays are unstoried, closely spaced, 2-4 seriate, essentially homogeneous to heterogeneous, 120 μm to 300 μm in height, and up to 40 μm in width.

COMMENTS

Juglans is distributed in the northwest Himalayas, the trans-Indus, regions, the Kuram valley, Sikkim and the hills of upper Burma and is cultivated in the Himalayas from 3,500 ft to 11,000 ft. Palaeobotanical and palynological evidence has revealed that *Juglans* has been growing in the valley since Pliocene-Pleistocene times (Lone 1987). However, *Juglans* today is mostly planted in the valley proper. It also occurs as a less frequent constituent of the forests at high altitude. Perhaps the tree was very widespread in the historical past. Once again, human hands might have led to its reduction. The inference of human interference leading to the reduction of oaks, birches, walnuts, and other trees from the valley can be justified by the fact that merely by removal of leaves for sheep a wholesale destruction of a birch forest at Kainmal, about 100 m above Gulmarg, has taken place in recent human memory (Vishnu-Mittre and Sharma 1966).

5.6.6. *Buxus* spp. (fig. 5.50)

The evidence of *Buxus* has come from Megalithic period at Burzahom.

The wood is diffuse porous. Growth rings are indistinct. Vessels are solitary or in pairs, oval to round, medium to thick-walled, and 20 μm to 60 μm in axial diameter. Parenchyma is apotracheal diffuse and sparse. Fibres are thin-walled to moderately thick-walled, oval, occasionally angular, and 10 μm to 18 μm in diameter. Rays are 1-2 seriate, normally spaced, unstoried, homogeneous, and up to 117 μm in height and 11 μm in width.

The main anatomical features of the wood are (1) diffuse porosity, (2) parenchyma apotracheal diffuse, and (3) rays 1-2 seriate. When these anatomical characters were compared with extant woods, it was found that the unknown resembles *Buxus* spp. in every respect (Miles 1978).

162   *Palaeoethnobotany*

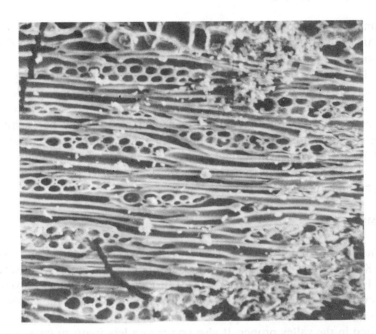

Figure 5.49: T.L.S. *Juglans* 200x.

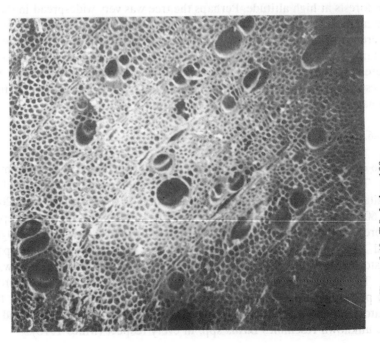

Figure 5.48: T.S. *Juglans* 100x.

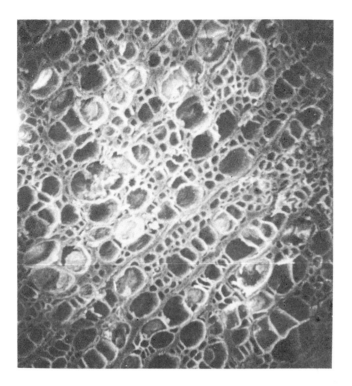

Figure 5.50: T.S. *Buxus* sp.

COMMENTS

Buxus spp. do not grow in the valley presently. Pleistocene deposits have revealed the occurrence of *B. wallichiana* and *B. papillosa* (Vishnu-Mittre 1965). *Buxus wallichiana* is growing in the neighbouring Jammu region and it seems probable that the wood of this tree has been transported from there.

5.6.7. *Ficus* spp. (figs. 5.51 and 5.52)

Use of *Ficus* is evident in Neolithic I phase at Burzahom and N.B.P. phase at Semthan.

The wood is semi-ring to diffuse porous. Growth rings are not distinct. Vessels are large, spherical to oval, thin-walled to medium thick-walled, usually in pairs or in groups of up to three occasionally solitary, and 50 μm to 125 μm in tangential diameter. Tyloses are abundant. Parenchyma is indistinct. Fibres are thin-walled to medium thick-walled, oval to oblong, and 15 μm to 20 μm in diameter. Rays are normally spaced, unstoried, 2-5 seriate, homogeneous, wide, and up to 250 μm in height.

COMMENTS

Species of *Ficus* are distributed in Punjab, the outer Himalayas, eastward to Nepal, and up to 5,500 ft on the hills of Marwa and Abu (Agarwal 1970, Brandis 1971). The tree is not an indigenous component of the present-day vegetation of the valley. However, *F. cunia* has been reported from Pleistocene deposits of the valley (Vishnu-Mittre 1984). The possibility of some trees growing near the sites is very remote and it is quite possible that *Ficus* has been introduced from outside probably for some religious purposes.

5.6.8. *Aesculus indica* (figs. 5.53 and 5.54)

Aesculus indica was recovered from all the four periods at Burzahom and Indo-Greek and Kushan periods at Semthan.

The wood is diffuse porous. Growth rings are not conspicuous. Vessels are small, numerous, nearly uniform in size, evenly distributed, spherical to oval, thin-walled, solitary or in multiples of two or more, and 20 μm to 45 μm in diameter. Vessel members are storied with other members as well as unstoried. Inter-vessel pits are orbicular to angular.

Parenchyma is apotracheal, scattered diffuse and scanty. Fibres are angular, thick-walled with small lumen, 5 μm to 15 μm in diameter. Rays are normally spaced, unstoried, homogeneous, uniseriate with occasional biseriation in some rays, up to 18 cells and 345 μm in height and 17 μm in width.

COMMENTS

Aesculus indica grows in the northwest Himalayas from 4,000 ft to 9,000 ft, the trans-Indus regions, in Kafiristan at 7000 ft to 8000 ft, and from the Indus to Nepal, chiefly in moist and shady valleys. It grows fairly in Kashmir forests and is cultivated as an ornamental plant in the valley plains. It has been a constant member of Pliocene, Pleistocene and post-glacial arboreal vegetation of the valley. Thus it must have been locally available.

5.6.9. *Acer* spp. (figs. 5.55 and 5.56)

The charcoals belonging to *Acer* spp. have been recorded from Hindu Rule period at Semthan.

The wood is semi-ring to diffuse porous. Growth rings are not conspicuous. Vessels are numerous, usually solitary, occasionally in pairs or

Archaeological Evidence

Figure 5.52: T.L.S. *Ficus* sp. 200x.

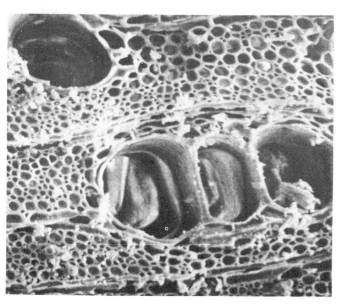

Figure 5.51: T.S. *Ficus* sp. 200x.

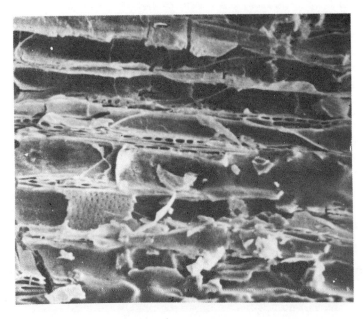

Figure 5.54: T.L.S. *Aesculus indica* 200x.

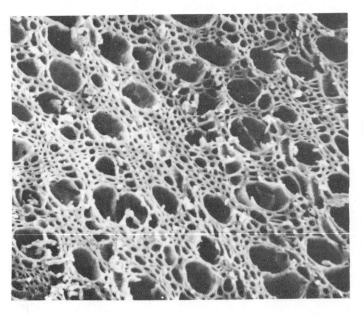

Figure 5.53: T.S. *Aesculus indica* 200x.

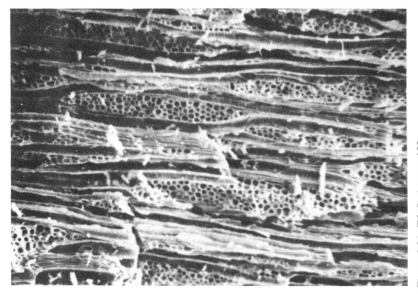

Figure 5.56: T.L.S. *Acer* sp. 100x.

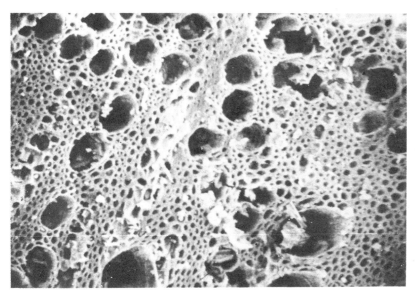

Figure 5.55: T.S. *Acer* sp. 200x.

multiples of three, oval to rounded, and 30 μm to 60 μm in tangential diameter. Inter-vessel pits are small and oval to angular. Parenchyma is sparse, paratracheal as well as apotracheal diffuse. Fibres are thin-walled to moderately thick-walled, oval to angular, and 5 μm to 12 μm in diameter. Rays are closely spaced, heterogeneous as well as homogeneous. Rays are of two types: narrow rays 1–3 seriate, 60 μm to 220 μm in height, 6 μm to 30 μm in width, and broad rays 3–7 seriate, 250 μm to 620 μm in height and 10 μm to 70 μm in width.

COMMENTS

The species of *Acer* commonly found in Kashmir are *A. oblongum*, *A. pentapomicum*, *A. villosum*, *A. caesium*, *A. caudatum*, and *A. pictum* (Singh and Kachroo 1976). *A. pentapomicum* grows from 2,300 ft to 2,700 ft, *A. caesium* from 4,000 ft to 10,000 ft, *A. villosum* at 7,000–9,000 ft, *A. caudatum* at 8,000–11,000 ft, *A. pictum* at 4,000–9,000 ft, and *A. oblongum* at up to 6,000 ft. Past record of the genus, as revealed by palaeobotanical evidence, goes back to Pleistocene times in the valley. Hence this genus, too, was locally available to the ancient inhabitants.

5.6.10. *Parrotiopsis jacquemontiana* **(figs. 5.57 and 5.58)**

Parrotiopsis jacquemontiana it is recorded at Semthan in Kushan and Hindu Rule phases.
The wood is diffuse porous. Growth rings are not conspicuous. Vessels are small, numerous, thin-walled to moderately thick-walled, almost of uniform size and evenly distributed, usually solitary and occasionally in groups of two or three, and 20 μm to 35 μm in tangential diameter. Inter-vessel pits are small, numerous, and oval to round. Parenchyma is metatracheal diffuse in the form of single-celled bands. Fibres are thick-walled, and 8 μm to 15 μm in diameter. Rays are fine, normally spaced, uniseriate to biseriate, occasionally 3-seriate, 90 μm to 260 μm in height and 8 μm to 30 μm in width, and homogeneous to heterogeneous.

COMMENTS

Parrotiopsis jacquemontiana grows in the Kuram valley, Kashmir, and Chamba from 3,800 ft to 8,500 ft. (Brandis 1971). It is well represented in the present-day vegetation of Kashmir and has been so since Pleistocene times.

5.6.11. *Salix* **spp. (figs. 5.59 and 5.60)**

Salix has been recovered from Indo-Greek phase at Semthan.

Archaeological Evidence

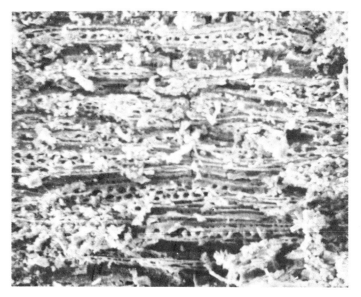

Figure 5.58: T.L.S. *Parrotiopsis jacquemontiana* 200x.

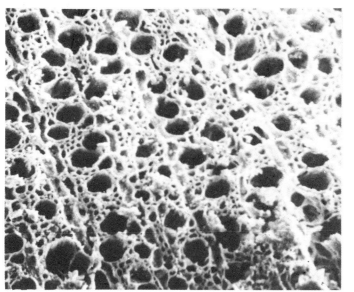

Figure 5.57: T.S. *Parrotiopsis jacquemontiana* 200x.

The wood is diffuse porous. Growth rings are conspicuous. Vessels are round to oval, usually solitary, and occasionally in multiples of two with tangential diameter of 20 μm to 40 μm Parenchyma is not distinct and is apparently terminal in position. Fibres are thin-walled to moderately thick-walled, angular, and 8 μm to 10 μm in diameter. Rays are fine, closely spaced, unstoried, uniseriate, occasionally bicelled in the middle, essentially heterogeneous, and up to 200 μm in height.

COMMENTS

Salix is distributed in the Kuram valley at 10,000 ft to 12,000 ft and is very common in the Himalayas from 7,000 ft to 8,000 ft. In Kashmir it grows at 6,000 ft to 8,000 ft. It has been recorded in the valley since Pliocene-Pleistocene times.

5.6.12. *Populus* spp. (figs. 5.61 and 5.62)

The poplar species have been identified from Neolithic I, Neolithic II, and Megalithic periods at Burzahom and Kushan phase at Semthan.
　　The wood is diffuse porous. Growth rings are not distinct. Vessels are oval to subglobular, usually in multiples of 2 to 5 or more, occasionally solitary, and 23 μm to 35 μm in diameter. Inter-vessel pits are oval to orbicular. Parenchyma is sparse in the form of diffuse cells. Fibres are thin-walled to medium thick-walled, angular to squarish, and 10 μm to 15 μm in diameter. Rays are unstoried, uniseriate, occasionally bicelled, essentially homogeneous, sometimes heterogeneous, 6 to 15 cells and 100 μm to 300 μm in height, and 10 μm to 15 μm in width.

COMMENTS

The common species of *Populus* in India are *P. euphratica*, *P. nigra*, *P. alba*, and *P. ciliata*. *Populus euphratica* is common in the forest belt of Sindh along the Indus. *Populus nigra* is frequently planted in the northwest Himalayas, particularly in Kashmir, and also in Ladakh as high as 12,500 ft. *Populus ciliata* is distributed in the northwestern Himalayas from 4,600 ft 10,000 ft, Sikkim from 3,500 ft to 9,000 ft, and in Bhutan (Troup 1921, Brandis 1971, Singh and Kachroo 1976).
　　Many species of *Populus* are known to have occurred in Kashmir since Pliocene-Pleistocene times (Vishnu-Mittre 1965, 1984, Lone 1987). Hence they were locally available to ancient people.

Archaeological Evidence 171

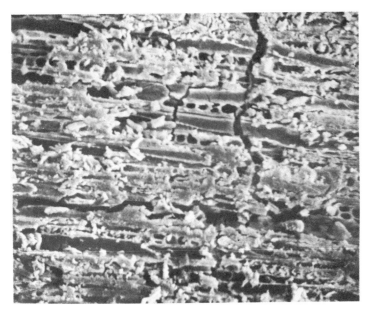

Figure 5.60: T.L.S. *Salix* sp. 200x.

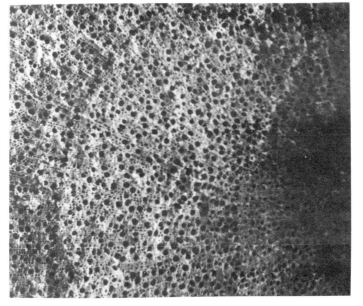

Figure 5.59: T.S. *Salix* sp. 50x.

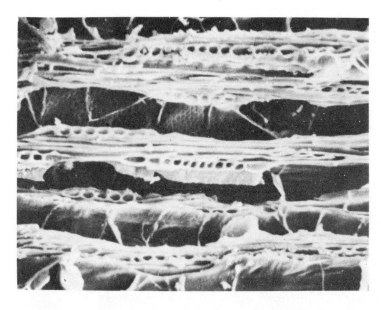

Figure 5.62: T.L.S. *Populus* sp. 200x.

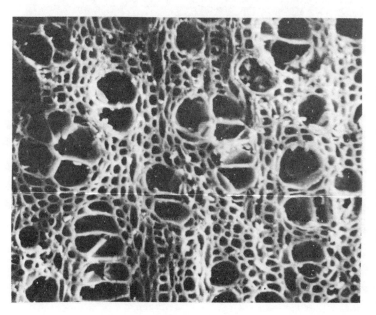

Figure 5.61: T.S. *Populus* sp. 200x.

5.6.13. *Crataegus oxyacantha* (figs. 5.63 and 5.64)

Crataegus oxyacantha has been recorded from Neolithic I period at Burzahom and Kushan and Hindu Rule phases at Semthan.

The wood is diffuse porous. Growth rings are not conspicuous. Vessels are medium-sized, numerous, thin-walled to moderately thick-walled, solitary or in pairs, 20 μm to 55 μm in tangential diameter, and oval to rounded. Tyloses are present. Inter-vessel pits are oval to round, small, and numerous. Parenchyma is apotracheal diffuse and scanty. Fibres are thick-walled, oval to angular with small lumen, and 6 μm to 10 μm in diameter. Rays are unstoried, 2–4 seriate, heterogeneous, 8 to 15 cells and 125 μm to 210 μm in height, and 30 μm to 40 μm in width.

COMMENTS

Crataegus oxyacantha is distributed in Baluchistan, the Kuram valley, the northwest Himalayas from the Indus to Ravi at 5,000 ft to 9,000 ft, Afghanistan, western Asia, Siberia, and Europe (Brandis 1971, Gamble 1972). It is fairly distributed in Kashmir valley forests. This genus is indigenous to the valley as evident in the palaeobotanical record.

5.6.14. *Pyrus pashia* (fig. 5.65)

The charocoals belonging to *Pyrus pashia* have been recovered from Semthan, Kushan phase.

The wood is semi-diffuse to diffuse porous. Growth rings are not distinct. Vessels are numerous, solitary or in multiples of up to five, oval to spherical, and 40 μm to 60 μm in tangential diameter. Inter-vessel pits are small and oval to orbicular. Parenchyma is apotracheal diffuse and scanty. Fibres are thin-walled to moderately thick-walled and 8 μm to 15 μm in diameter. Rays are normally spaced, unstoried, essentially homogeneous, 2–4 seriate, and up to 240 μm in height and 25 μm in width.

COMMENTS

Pyrus pashia grows in Afghanistan, the trans-Indus regions, the Himalayas from Hazara to Bhutan at 2,500 ft to 8000 ft, the Khasi hills, Manipur, Upper Burma, and from Kashmir to Kumaon. It is an Indigenous plant locally available in the past.

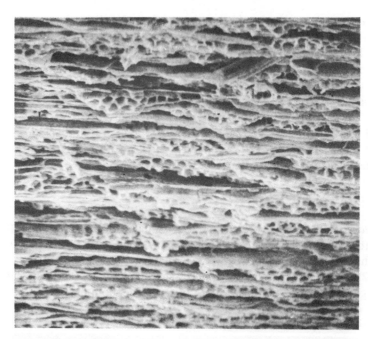

Figure 5.64: T.L.S. *Crataegus oxyacantha* 200x.

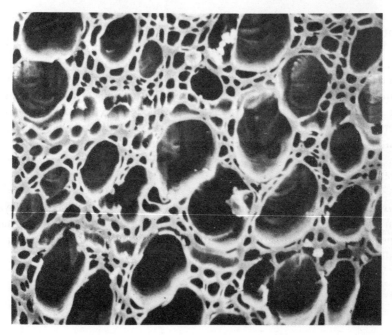

Figure 5.63: T.S. *Crataegus oxyacantha* 200x.

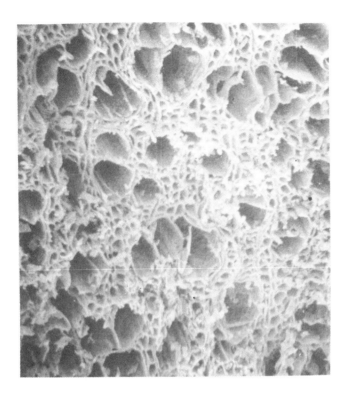

Figure 5.65: T.S. *Pyrus* sp. 200x.

5.6.15. *Fraxinus excelsior* (figs. 5.66 and 5.67)

Fraxinus excelsior was recorded from Neolithic II and Post-Megalithic periods at Burzahom and N.B.P. and Kushan phases at Semthan.

The wood is ring porous. Growth rings are distinct. Vessels are large, solitary or in groups, medium-sized, rounded forming continuous bands, and 30 μm to 75 μm in tangential diameter. Inter-vessel pits are oval to rounded. Parenchyma is paratracheal, forming a sheath. Fibres are numerous, thin-walled to medium thick-walled, angular to round, and 12 μm to 17 μm in diameter.

Rays are normally spaced, unstoried, homogeneous, 2–5 seriate, 9 to 20 cells and 140 μm to 265 μm in height and 27 μm to 60 μm in width.

COMMENTS

Fraxinus excelsior is distributed over the northwestern Himalayas, in the basin of the Jhelum, Chenab, and Ravi rivers from 4,000 ft to 9,000 ft, in

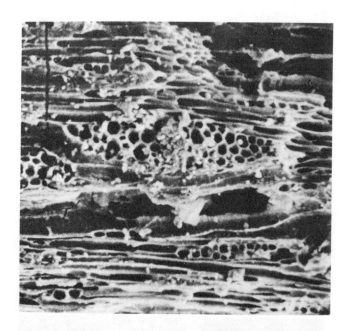

Figure 5.67: T.S. *Fraxinus excelsior* 200x.

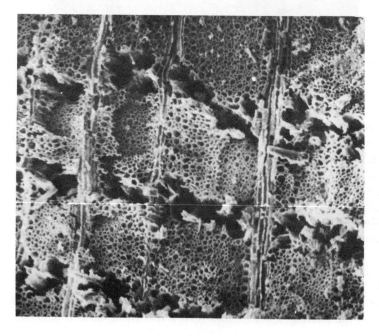

Figure 5.66: *Fraxinus excelsior* 200x.

Europe, and in the mountains of western Asia (Brandis 1971). As revealed by its past record in the valley, it is indigenous and must have been available locally.

5.6.16. *Ulmus wallichiana* (figs. 5.68 and 5.69)

Elm has been recorded at Burzahom in the Neolithic I period and at Semthan in the N.B.P. and Indo-Greek periods.

The wood is ring porous. Early wood vessels are large and arranged in more or less continuous wavy concentric bands. Vessels are numerous, either solitary or in multiples of 2 to 3, and ovoid to round. Transition from early to late wood is abrupt. Late wood pores are small, numerous, clustered, and 30 μm to 80 μm in diameter, and early wood pores are 70 μm to 150 μm in diameter. Inter-vessel pits are small, oval to orbicular, and 2 μm to 10 μm in diameter. Parenchyma is paratracheal, scanty, and vesicentric. Fibres are numerous, thin-walled to thick-walled, and 5 μm to 10 μm in diameter. Rays are wide, unstoried, 3–8 seriate, essentially homogeneous, and up to 600 μm in height and 60 μm in width.

COMMENTS

Ulmus wallichiana is distributed over the northwestern Himalayas from the Indus to Nepal at 3,500 ft to 10,000 ft, and is well represented in the Kashmir valley. It is of prolific distribution in the fossil record and is indigenous.

5.6.17. *Celtis australis* (figs. 5.70 and 5.71)

Post-Megalithic period at Burzahom and Pre-N.B.P. and Kushan periods at Semthan have revealed charcoals of *Celtis australis*.

The wood is semi-ring to ring porous. Growth rings are distinct. Early wood vessels are larger than the late wood vessels. Vessels in general are rounded, solitary, and 70 μm to 150 μm (average 120 μm) in diameter. Occasionally tyloses are present. Inter-vessel pits are small, and oval to orbicular. Parenchyma is paratracheal scanty to vesicentric. Fibres are angular, thick-walled, 8 μm to 12 μm in diameter. Rays are narrow to moderately wide, normally spaced, unstoried, 1–4 seriate and heterocellular, 110 μm to 320 μm and 4 to 16 cells in height and up to 38 μm in width.

COMMENTS

Celtis australis is an indigenous tree species of Kashmir and is very common along the graveyards in the valley.

Figure 5.69: T.L.S. *Ulmus wallichiana* 200x.

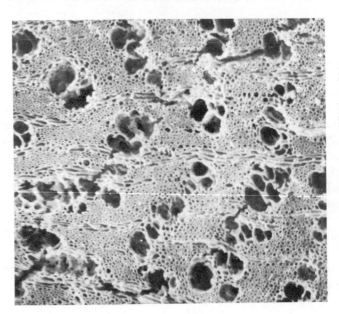

Figure 5.68: T.S. *Ulmus wallichiana* 200x.

Archaeological Evidence

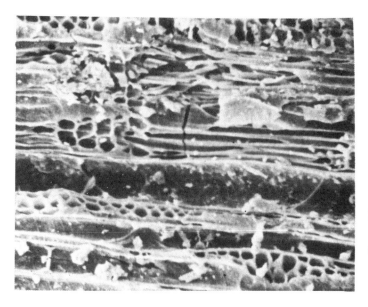

Figure 5.71: T.L.S. *Celtis australis* 200x.

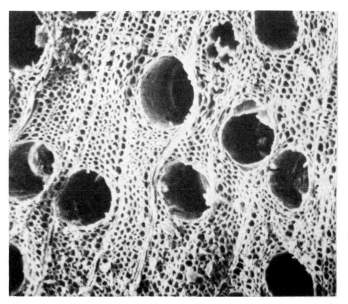

Figure 5.70: T.S. *Celtis australis* 200x.

5.6.18. *Prunus* spp. (fig. 5.72)

Prunus spp. were recorded at Semthan from N.B.P. and Kushan periods.

The wood is semi-ring porous. Growth rings are distinct. Vessels are medium-sized, those at the beginning of the ring somewhat large, closely placed, and aligned in more or less a continuous row, and those elsewhere in the ring quite uniform in size and evenly distributed. Vessels are solitary or in multiples of up to 4, and 35 μm to 80 μm in tangential diameter. Inter-vessel pits are oval to orbicular.

Parenchyma is very scanty. Fibres are angular, thin-walled to thick-walled, and 8 μm to 15 μm in diameter. Rays are 2–4 seriate, unstoried, homogeneous to heterogeneous.

COMMENTS

Some species of *Prunus*, such as *P. cerasus*, *P. domestica*, and *P. cerasifera*, are growing wild in the valley forests and are also represented in the fossil record. The species may have been locally available.

5.6.19. *Viburnum* spp. (fig. 5.73)

Viburnum is recorded at Semthan from the Indo-Greek phase.

The wood is diffuse porous. Growth rings are not distinct. Vessels are oval, rounded to angular, usually solitary and occasionally in pairs, and 20 μm to 40 μm in tangential diameter.

Parenchyma is sparse, not well defined. Fibres are thin-walled to moderately thick-walled, 8 μm to 16 μm in diameter. Rays are closely spaced, uniseriate, occasionally bicelled in the middle, and heterogeneous.

COMMENTS

Viburnum is a very common deciduous shrub in the forests of Kashmir along the slopes from 5,200 ft to 7000 ft. Palaeobotanical records of the genus are available since Pliocene times.

5.6.20. *Vitis vinifera* (figs. 5.74 and 5.75)

Vitis wood is recorded from Neolithic II period at Burzahom.

The wood is diffuse porous. Growth rings are indistinct. Vessels are round to oval, solitary or in groups of two, and 30 μm to 80 μm in diameter. Parenchyma is paratracheal and scanty. Fibres are angular with thin to thick walls and 8 μm to 15 μm in diameter. Rays are normally spaced, 3–5 seriate, unstoried, essentially homogeneous, up to 300 μm in height and 50 μm in width.

Archaeological Evidence 181

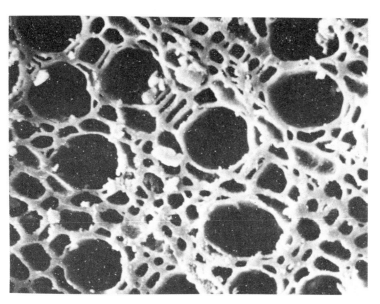

Figure 5.73: T.S. *Viburnum* sp. 500x.

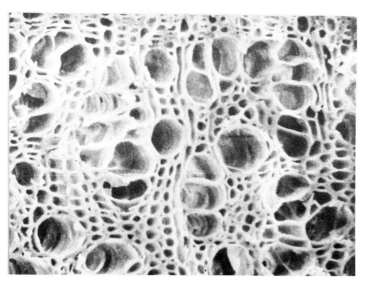

Figure 5.72: T.S. *Prunus* sp. 500x.

182 Palaeoethnobotany

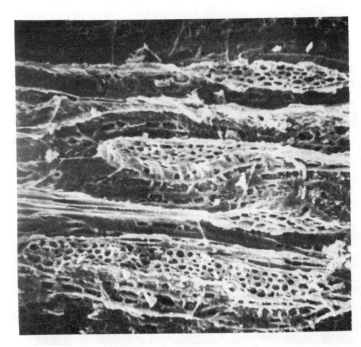

Figure 5.75: T.L.S. *Vitis vinifera* 200x.

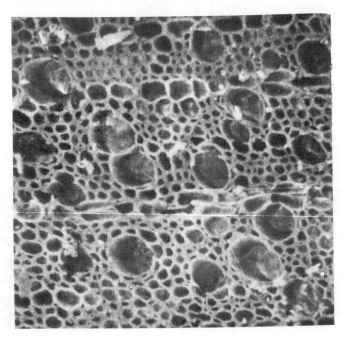

Figure 5.74: T.S. *Vitis vinifera* 200x.

5.6.21. Leguminous woods (figs. 5.76 and 5.77)

Woods identifiable as belonging to Leguminosae are recorded at Burzahom in Neolithic II and Megalithic periods and at Semthan in Pre-N.B.P. phase.

The wood is diffuse porous. Growth rings are indistinct. Vessels are rounded to oval, solitary or in pairs, medium-sized, thick-walled, 30 μm to 75 μm in diameter. Parenchyma is paratracheal, not forming a sheath. Fibres are abundant, circular to angular, thick-walled, 5 μm to 15 μm in diameter. Rays are widely spaced, unstoried, homogeneous to heterogeneous, 2–6 seriate, 200 μm to 350 μm in height, and up to 50 μm in width.

COMMENTS

Sen (1943) has remarked that comparative study of wood anatomy produced 'no sharp lines separating various subfamilies and genera'. Hence the material could not easily be assigned to any particular member.

5.6.22. *Pinus wallichiana* (figs. 5.78 and 5.79)

Neolithic I and Megalithic periods at Burzahom and Pre-N.B.P. and N.B.P. periods at Semthan revealed the presence of charcoals identifiable as *Pinus wallichiana*.

The wood is non-porous. Growth rings are distinct due to demarcation between early wood and late wood tracheids. Early wood tracheids have thinner walls and larger lumen (20 μm to 30 μm) than late wood tracheids, which have thicker walls and smaller lumen (10 μm to 18 μm). Tracheids are arranged in definite radial rows and are squarish to polygonal. Vertical resin canals are present, mostly circular, and 120 μm to 170 μm in diameter. Epithelial cells surrounding the resin canals are crushed, and thin-walled. Vertical parenchyma is lacking. Rays in cross-section appear to be uniseriate. Pits on the radial walls are large, bordered in single rows.

Ray tracheids have wavy walls. Cross field pits are large, window-like, 1 to 2 in each cross field.

COMMENTS

Pinus wallichiana is presently distributed over the Kuram valley at 8,000 ft to 11,000 ft, Safed Koh, Kafiristan, the Himalayas and also in more arid valleys such as Lahaul and Kunawar, eastward as far as Nepal. It also grows in Bhutan and Afghanistan (Troup 1921, Raizada and Sahni 1960, Brandis 1971, Gamble 1972).

Palaeobotany and palynology of the Lower Karewas and post-glacial

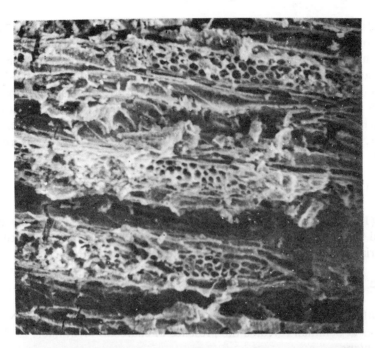

Figure 5.77: T.L.S. Leguminous wood 200x.

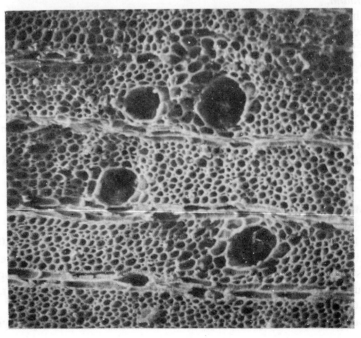

Figure 5.76: T.S. Leguminous wood 200x.

Archaeological Evidence

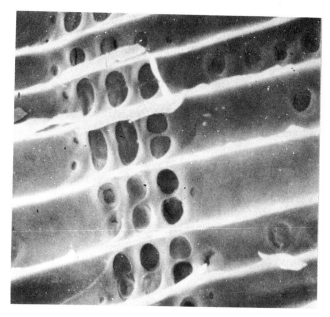

Figure 5.79: R.L.S. *Pinus wallichiana* 500x.

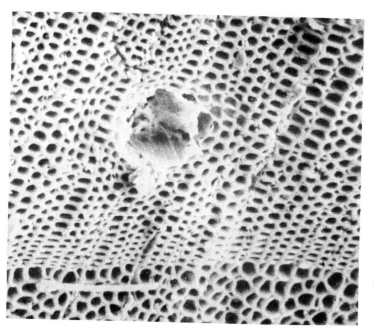

Figure 5.78: T.S. *Pinus wallichiana* 200x.

deposits in the Kashmir valley (de Terra and Paterson 1939, Puri 1948, Vishnu-Mettre et al. 1962, Vishnu-Mittre 1965, 1984, Gupta et al. 1985, Sharma et al. 1985, Sharma and Gupta 1985, Lone 1987, Lone et al. 1988) have revealed that *Pinus wallichiana* has been growing continuously in Kashmir since Pliocene-Pleistocene times. Therefore, the wood must have been locally available to the ancient inhabitants.

5.6.23. *Picea smithiana* (figs. 5.80 and 5.81)

Neolithic II phase at Burzahom and Pre-N.B.P. and Indo-Greek phases at Semthan have revealed charcoals of *Picea smithiana*.

The wood is non-porous. The bulk of the wood is made up of tracheids arranged in definite radial rows. Tracheids are squarish to polygonal and 12 μm to 35 μm (average 22 μm) in diameter. Vertical parenchyma cells are absent. Vertical resin canals are present, 70 μm to 150 μm in diameter. Epithelial cells surrounding the resin canals are crushed and thick-walled. In tangential section are seen two types of rays: the uniseriate rays 2 to 10 cells and 60 to 260 μm in height and the fusiform rays which possess horizontal resin canals up to 5 cells and 45 μm in width and 15 to 19 cells in height. The horizontal resin canals are 60 μm to 90 μm in diameter. Tracheids possess bordered pits on the tangential walls, and spiral thickenings appear on the radial walls of longitudinal tracheids. The cross field pits are of piceiod type.

COMMENTS

Presently *Picea smithiana* grows in the Kuram valley from 8,000 ft to 12,000 ft, Kafiristan, Chitral, Gilgit, the Himalayas from Kashmir to Garhwal from 7,000 ft to 11,000 ft, and the inner valleys of Sikkim and Bhutan from 8,000 ft to 15,000 ft (Brandis 1971, Gamble 1972). In the Kashmir valley forests *Picea* is less abundant compared to *Pinus* and *Abies*.

Palaeobotanical and palynological investigations of Karewa and postglacial sediments (de Tura and Paterson 1939, Puri (1945) Vishnu Mittre et al. (1962) Vishnu Mmittre (1965, 1984, Gupta et al. 1985, Sharma et al. 1985 Sharma and Gupta 1985, Lone 1927, Lone et al. 1988), indicate the existence of *Picea* in Kashmir for the past four million years or so. Hence it was locally available to the ancient inhabitants.

5.6.24. *Abies pindrow* (fig. 5.82)

Abies pindrow occurred in the N.B.P. phase at Semthan.

The wood is non-porous. Growth rings are distinct. Tracheids are squarish to polygonal and aligned in definite radial rows, and the last 4 to 5 rows of the late wood are tangentially flattened. Transition from early to late

Archaeological Evidence 187

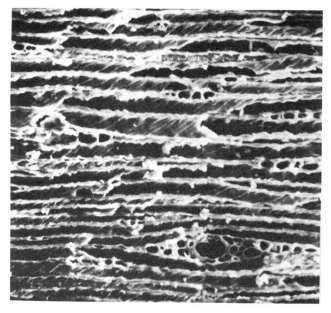

Figure 5.81: T.L.S. *Picea smithiana* 200x.

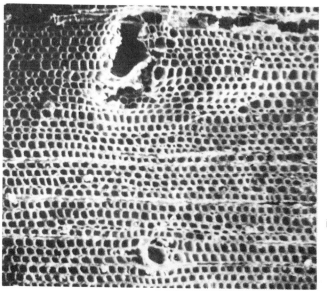

Figure 5.80: T.S. *Picea smithiana* 200x.

188 Palaeoethnobotany

wood is gradual. Early wood tracheids have large lumen (20 μm to 30 μm) and the late wood tracheids have smaller lumen (10 μm to 25 μm). Vertical parenchyma, resin, and traumatic canals are lacking. Rays are uniseriate and 5 to 10 cells and 60 μm to 90 μm in height. Spiral thickenings are seen on the radial walls of the tracheids. Cross field pitting is of taxodioid type.

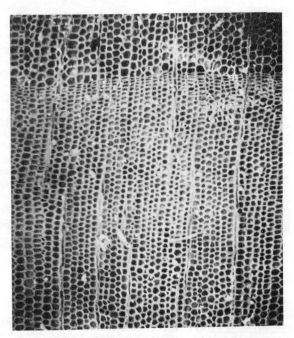

Figure 5.82: T.S. *Abies pindrow* 100x.

COMMENTS

The present distribution of *Abies pindrow* in India is in the Kuram valley from 8,000 ft to 11,000 ft, Chitral, and the outer Himalayas from 8,000 ft to 10,000 ft (Raizada and Sahni 1960, Brandis 1971, Gamble 1972). It is fairly distributed in the Kashmir valley. Palaeobotanical and palynological studies indicate its existence in the valley since Pliocene times. It was locally available to the ancient people.

5.6.25. Cedrus deodara (figs. 5.83 and 5.84)

Cedrus deodara has been recovered from the post-Megalithic period at Burzahom and Pre-N.B.P. and N.B.P. periods at Semthan.

Archaeological Evidence

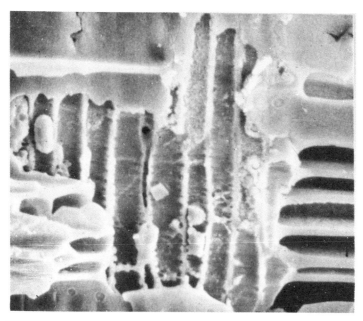

Figure 5.84: R.L.S. *Cedrus deodara* 500x.

Figure 5.83: T.S. *Cedrus deodara* 200x.

The wood is non-porous. Growth rings are distinct. Tracheids are arranged in radial rows. Early wood tracheids are polygonal and 30 μm to 55 μm in diameter, late wood tracheids are tangentially flattened and 15 μm to 30 μm in diameter. Transition from early to late wood is slightly abrupt. Vertical parenchyma is absent. Resin canals are lacking but traumatic canals are present, 70 μm to 80 μm in tangential diameter. Rays in cross-section appear to be uniseriate. Radial walls of the tracheids possess large, scalloped bordered pits. Ray tracheid walls are wavy. Cross field pitting is of taxodioid type.

COMMENTS

Cedrus deodara is distributed in Afghanistan, the Kuram valley from 7,500 ft to 10,000 ft, Chitral, the northwestern Himalayas at 4,000 ft to 10,000 ft ascending at places up to 12,000 ft, and the basin of the principal tributaries of the Indus, Tous, Jumna and Bhagirati rivers. It is cultivated in Kumaon and Nepal (Troup 1921, Pearson and Brown 1932, Raizada and Sahni 1960, Brandis 1971, Gamble 1972). In Kashmir, though it is almost absent on the Pir Panjal side, it is found in other parts of the valley both in dry and moist regions. It is of limited occurrence, unlike *Pinus* and *Abies*. It has been growing in Kashmir since Plio-Pleistocene times (Lone 1987) and was locally available to ancient inhabitants.

5.6.26. *Cupressus* spp. (figs. 5.85 and 5.86)

Charcoals belonging to *Cupressus* spp. are identified from Kushan and Hindu Rule phases at Semthan.

The wood is non-porous. Growth rings are indistinct. The bulk of the wood is made up of tracheids, aligned in definite radial rows. Tracheids are oval, rounded to polygonal and 15 μm to 28 μm in diameter. Resin canals are absent. Parenchyma cells are discernible in cross-section, diffused in between the tracheids. Tracheids possess bordered pits on their tangential and radial walls. Ray tracheids are wanting. Rays are uniseriate, 3 to 5 cells and 25 μm to 87 μm in height, and 5 μm to 8 μm in width. Cross field pits are cupressoid.

Among the conifers, the recovery of *Cupressus* spp. from archaeological contexts is the most interesting one. There is no indigenous species of the genus in Kashmir. *Cupressus* spp. grow on the high mountains on the outer ranges of the Himalayas from Chamber to Nepal at 6,000 ft to 9,000 ft (Brandis 1971, Gamble 1972). From the pollen analysis of some Lower Karewa deposits pollen belonging to *Cupressus* have been reported (Gupta et al. 1985, Sharma et al. 1985), Sharma and Gupta 1985). This leads us to think that the Semthan charcoals belong either to the Pleistocene remnants

Archaeological Evidence 191

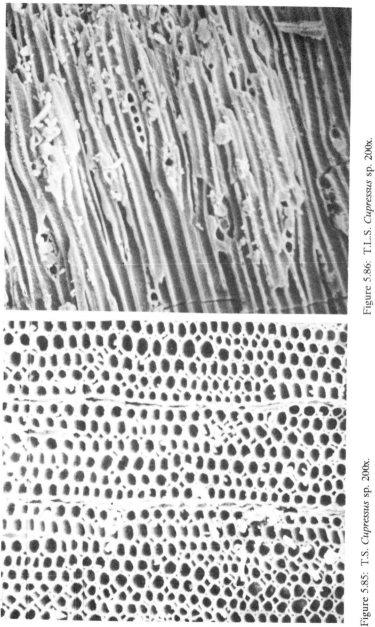

Figure 5.86: T.L.S. *Cupressus* sp. 200x.

Figure 5.85: T.S. *Cupressus* sp. 200x.

of *Cupressus* in the valley or to the trees introduced during period IV at Semthan from outside. In the light of the present distribution, it appears that *Cupressus* was introduced into the valley from the area of its distribution, probably in the Punjab Himalayas, through the contact of the inhabitants. The identification of *Cupressus* pollen in the Pleistocene deposits needs to be re-examined. If the pollen are really of *Cupressus*, then they might have come from much further away from the Punjab Himalayas. So far no megafossils of *Cupressus* have been recovered from the Pleistocene deposits of the valley.

CHAPTER 6

DIFFUSION OF PLANTS INTO KASHMIR

The preceding chapter makes it clear that there are only a few plants that have been available to the ancient inhabitants of Kashmir locally. Those that were available mostly catered to the need for wood. Ethnobotanically speaking, the way plants are used by peoples of different regions indicates both their development and the nature of culture and also the origin and history of the cultivated plants they use (Li 1970). Man uses plants first for food, next for clothing, and later for industrial and other purposes. We have already seen that none of the food plants used by ancient man in Kashmir evolved locally, except perhaps *Triticum sphaerococcum*. Almost all were brought from their places of origin. The plants have been brought from the Indian plains, China, West Asia, Central Asia, the Mediterranean region, and other places. In order to know about the diffusion of various plants it seems imperative to look for the ancient routes which, if they were available, were followed by ancient people to bring plants into the valley. Such routes were available since Palaeolithic times (Dikshit 1982, Kachroo 1986).

Sahni (1936) has stated that 'round about Middle Pleistocene time when the main valley of Kashmir was still occupied by the great "Karewa Lake", interglacial man of about the same stage of cultural development as Neanderthal or Mousterian man in Europe and as Peking man in the Far East flourished (a) in the plains of northern Punjab; (b) on the shores of Karewa lake in the heart of Kashmir; and (c) just across the Great Himalayan range indicating contact between early human cultures on the two sides of the main Himalaya and the Pir Panjal range.' The discovery of Palaeolithic stone flake industries in three widely separated parts of northern India, southwest of Rawalpindi (Chitta, Pakistan), a few kilometres east of Srinagar (Pampur, Kashmir), and at Kargil (Ladakh), reveal with respect to human activities. Kargil lies beyond the main Himalayan range on the ancient trade route over Zoji-la connecting India with Central Asia, Tibet, and China. In this connection de Terra's remark (de Terra's and Paterson

1939) is significant that traces of prehistoric human industry have been discovered even north of the central Himalayan range on the borders of Little Tibet. In fact, Sahni (1936) asserted that the Himalayas and the Pir Panjal range could not act 'as a barrier to the migration of Palaeolithic or even Neolithic man.' Thus, long before man conquered the ocean, intercourse between the ancient cultures of India and China was possible by the direct route across the Himalayas even in Palaeolithic and Neolithic times.

The circumstantial evidence that presents itself is that during Neolithic times, from a nucleus in Central Asia there could have been a two-fold migration, one to the southwest and thence to Kashmir and one northeast to China, Manchuria, and Siberia. Dikshit (1982) has also stated that 'the entry to Kashmir was through Gilgit and Sarhad and then along the foot-hills of Kun Lun ranges.' But has Neolithic man trod these routes into Kashmir? There is no 100% proof, but there are historical facts and cultural links.

The roads carved by man since prehistoric times played an important role in the dissemination of one people's culture to the other. The most important ancient 'highway' was the silk route (fig. 6.1), which connected China with Europe. West Asia, and India. Other roads that connected India with Central Asia passed through its north and northeastern regions. Central Asia was also connected with Tibet and with the Kashmir valley through the northeastern regions and these were connected with the silk route (Ahmad 1986).

There were three main roads, linking Kashmir with Central Asia. One lay in the northwest, passing through Baramulla to Muzaffarabad. Roads also led from the valley to Mansura near Hyderabad (Pakistan) and it is interesting to note that Kashmir merchants sailed down the Indus from the Jhelum as far as the delta of the Indus. Another road that passed from the valley to Central Asia was via Gilgit and Chitral, and the important road that connected the valley with Tibet and Sinkiang passed through Leh and the Karakorum mountains (Ahmad 1986).

Any of these routes could have been followed by the ancient inhabitants in bringing plants like *Prunus persica, P. armeniaca, P. amygdalus, Panicum*, and *Morus alba* from China and Central Asia. As for the crops of Near Eastern origin like wheat, barley, peas and lentil, these could have been brought through Harappa or the Indo-Gangetic plains. There are resemblance and indications of contact between the Harappan and Burzahom Neolithic cultures in respect to pre-Harappan pot and semi-precious stones found in Burzahom (Buth and Kaw 1985) on the one hand and the pre-N.B.P. phase at Semthan and late Harappan Bara pottery at Banawali in Haryana (Bisht 1986) on the other. Sharma (1982) has also suggested Harappan contact in the late phase of Neolithic at Gofkral. Further, an idea has been put forth that Neolithic culture of Central Asia may be of West Asian origin (Dikshit 1982, Kachroo 1986) and therefore crops might have been passed to Kash-

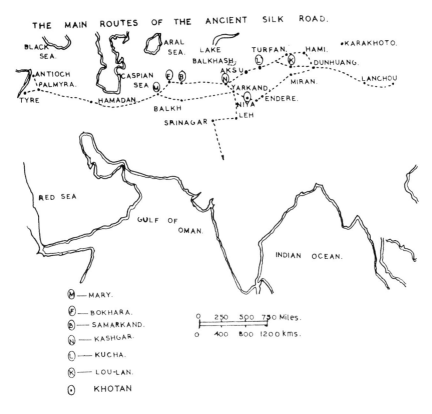

Figure 6.1: Map showing main routes of the ancient silk route.

mir via central Asia. Similarly, rice was also carried from the Indo-Gangetic plains (Buth et al. 1986a), as might have been *Phaseolus* spp. and *Ficus* spp. *Platanus orientalis*, being of Mediterranean origin, might have been brought via Central Asia.

Taking into consideration all the pros and cons of diffusion processes (Harlan 1986, Zohary 1986), we have come to the conclusion that the plant species used by ancient man in Kashmir can be grouped into the following categories:

1. *Plants diffused from West Asia*: These include *Triticum aestivum, Hordeum vulgare, Pisum sativum,* and *Lens culinaris*. Possibly they were introduced via Harappan cultures.

2. *Plants diffused through Central Asia*: These include the temperate horticultural crops *Prunus amygdalus, P. persica* and *P. armeniaca*, ornamental *Platanus orientalis, Morus alba,* and probably *Panicum* and *Vitis vinifera*.

3. *Plants diffused through Punjab and the northwestern Himalayas*: These

include the cereals *Oryza sativa* and *Triticum sphaerococcum* and the woody plants *Buxus* and *Ficus*.

4. *Locally available plants*: These include *Juglans regia* among the fruit crops, all the woods recovered except those mentioned above, and all the fodder plants (weed seeds).

CHAPTER 7

PROBABLE USES OF THE PLANTS RECOVERED

In the absence of any artefacts and from the nature of plant remains one can at best speculate on the uses to which various plant species might have been put by the inhabitants in the light of their present-day uses. The uses might have been as follows:

Oryza sativa: Food after cooking in water; buns and bread made from flour; paddy and broken rice as poultry feed; husk as fuel and binding material; culms as cattle feed; grass for roof thatching.

Triticum aestivum and *T. sphaerococcum*: Food in the form of bread and porridge; animal feed; fresh culms as fodder; residues left after threshing and grinding as cattle feed.

Hordeum vulgare: For making bread and porridge; fodder and cattle feed; probably for making some crude liquor.

Avena fatua and *A. sativa*: For livestock feed, especially winter feed for stalled animals.

Panicum spp.: As meal, bread or porridge especially in drought and scarcity; cattle feed.

Setaria spp.: As meal, bread or porridge, especially in drought and scarcity; cattle feed.

Phaseolus aureus: As dhal and vegetable; haulms as fodder; hulls and split beans as livestock food; hay as green manure.

P. mungo: Green pods as vegetable; dhal; hulls and straw as cattle feed; green manure.

P. aconitifolius: As human and animal food; forage; hay for livestock; green manure.

Pisum sativum: As vegetable; human and stock feed; green manure.

Lens culinaris: As dhal; husks, bran and dried haulms as fodder for livestock.

Juglans regia: Kernels eaten fresh or dried; oil extracted from kernels;

stones used as fuel; wood used for making household items, furniture; firewood.

Prunus persica: As edible fruit; firewood and fuel; cash crop.

P. cerasus: Edible fruit; fuel.

P. armeniaca: Edible fruit; oil; fuel; firewood.

Weed seeds: For feeding livestock.

Celtis australis: Firewood; to a lesser extent for agricultural and household items; leaves as fodder; sacred tree.

Pinus wallichiana: Building constructions and houseposts; agricultural and household items; firewood.

Picea smithiana: Building material; household items; firewood.

Cedrus deodara: Building construction; boat making; agricultural implements; household items; firewood.

Abies pindrow: Construction purposes; agricultural implements; firewood.

Cupressus spp.: Ornamental; firewood.

Aesculus indica: Firewood; household items.

Betula utilis: Bark for roof thatching; thin papery layers as writing material; firewood.

Ulmus wallichiana: Construction purposes; firewood, leaves as fodder.

Quercus spp.: Building material; furniture; agricultural implements; firewood; charcoal; leaves as fodder.

Fraxinus excelsior: Agricultural implements; furniture; firewood; fodder.

Ficus spp.: For some religious ritual.

Populus spp.: Construction purposes; agricultural and household implements; firewood; fodder.

Salix spp.: Agricultural and household items; furniture; firewood; fodder; twigs chewed for cleaning teeth.

Viburnum spp.: Firewood; fencing.

Parrotiopsis jacquemontiana: Agricultural implements; firewood; fencing.

Crataegus oxyacantha: Agricultural implements; firewood.

Pyrus spp.: Edible fruit; fodder; fuel.

Acer spp.: Furniture; building material; firewood.

Morus alba: Silkworm rearing; agricultural implements; furniture; fodder; firewood.

Platanus orientalis: Ornamental; furniture; firewood.

CHAPTER 8

VEGETATION, CLIMATE AND THE BIOTIC FACTOR

Apart from their archaeological and botanical interest the identification of plant remains is of considerable significance in view of the light they throw on past vegetation and the climate that prevailed in the region during the period of occupation. The present investigations indicate that the coniferous elements in the vegetation included *Pinus wallichiana, Picea smithiana, Abies pindrow*, and *Cedrus deodara*. Broad-leaved trees included *Quercus* spp., *Ulmus wallichiana, Fraxinus excelsior, Celtis australis, Betula utilis, Populus* spp., *Aesculus indica, Juglans* spp., *Acer caesium, Pyrus pashia, Prunus* spp., and shrubs like *Salix wallichiana, Parrotiopsis jacquemontiana*, and *Viburnum* spp. The introduced elements included *Cupressus* spp., *Buxus, Platanus orientalis, Morus alba*, and *Ficus* spp. All these give an indication of arboreal vegetation. Some evidence of ground vegetation has come through the identification of weed seeds like *Lithospermum arvense, Galium tricorne, G. asperuloides, G. aparine, Melilotus albus, Medicago* spp., *Ipomoea* spp., *Astragalus* spp., and *Lathyrus Vicia* spp. The agroecosystems consisted of rice in the irrigated fields and wheat, barley, oats, and pulses in the drylands. Some horticultural trees like *Juglans* and *Prunus* were also probably cultivated. It would be interesting to find out how this compares with the present vegetation of the valley.

Except *Quercus, Buxus, Ficus, Platanus*, and *Cupressus*, all the species grow as indigenous components of the present-day vegetation. Even exotic species of these trees, like *Quercus robur, Ficus palmata, Cupressus sempervirens*, and *Platanus orientalis*, are grown and regenerate in the valley. Does this mean that these trees were growing in the valley three to five thousand years ago? Of course *Quercus* is the most important of all. Its charcoals are not found after period II at Semthan, i.e., 2200 B.P. The tree might have been growing in very low frequencies in the forests and as a result of biotic factor and some still unknown reasons, it was reduced to omission.

Further, it can be speculated that *Pinus, Cedrus, Picea, Abies, Acer, Crataegus, Parrotiopsis*, and *Viburnum* were restricted only to the forests, whereas *Salix, Populus, Juglans, Ulmus, Celtis, Prunus, Pyrus*, and *Fraxinus* could also have been propagated by the inhabitants in their own locality.

The frequency and number of charcoals can hardly give an idea about the relative frequency of different trees. From a comparative study of the past flora, as revealed by the plant remains, and the present vegetation, it seems reasonable to conclude that forest cover of the region on the whole has remained more or less of the same type. The floristic complex indicated by the charcoal determinations is generally characteristic of a temperate forest (Champion and Seth 1968). Taking all these factors into consideration, we would not be far wrong in assuming that the climate and rainfall of the valley have not changed to any appreciable extent during last five thousand years or so, except for the changes induced by man.

The direct or indirect anthropogenic influence of man on vegetation have resulted from (1) disturbance of the natural vegetation (clearing of forests for agriculture, timber, fuelwood, fodder); (2) transformation of the natural landscape into agricultural and pastoral land; (3) destruction of indigenous species through transformations such as silvicultural practices, which tend to favour a few species at the expense of many others; and (4) introduction and spread of exotics.

The clearance of forests for agricultural purposes is revealed by the palynological evidence of lake sediments of the valley which could be dated to around 4000 years B.P. (Vishnu-Mittre and Sharma 1966; Dodia et al. 1985). This evidence is corroborated by the retrieval of cereal grains belonging to wheat and barley at Burzahom giving a ^{14}C date of 4325 years B.P.

Agriculture is the main cause of transformation of the natural landscape, but agriculture apart man has favoured and propagated non-edible, useful species by raising them in populations. The common plantations raised in Kashmir under modern social forestry programmes are those of *Salix* spp., *Populus* spp., *Ailanthus* spp., and *Robinia pseudoacacia*. Besides, avenue trees along roadsides include *Platanus orientalis, Aesculus indica*, and poplar species (*Populus deltoides, P. nigra, P. alba*, and *P. tremula*).

As far as the introduction of exotics into the valley is concerned, Kashmir has had relations with the exterior world from very ancient times. The Burzahom culture dates the links between Kashmir and Central Asia to 4300 years B.P. Similarly, Kashmir was connected with the rest of India through passes in the Pir Panjal and with China and other neighbouring countries through the famous silk route and many other ancient routes (Ahmad 1986). Exchange of people and culture has resulted in the introduction of tree species such as *Platanus orientalis, Ficus* spp., *Morus* spp., and *Buxus* spp. at different stages of human occupation. Several plant species were introduced by the Moghals and later by the British and some of these, such as *Sternbergia* and *Ixiolirion*, have become naturalized here (Vishnu-Mittre 1966). The afforestation in dry, temperate hill slopes of Shankaracharya Hill in the valley (Firdous 1944, Mehta 1947) has revealed that subtropical taxa such as *Ailanthus glandulosa* and *Melia azedarech* together with exotic taxa like *Pinus helepensis, P. canariensis,* and *Cupressus arizonica* obtained here find the climatic conditions quite suitable.

As many as seven species of *Populus* grow in the valley, of which *P. nigra, P. alba, P. deltoides, P. euphratica* and *P. tremula* are planted mostly from cuttings and suckers. The woodland poplar, *P. ciliata*, is very sporadic and limited in Kashmir. There are several species of *Salix*, of which *S. oxycarpa, S. lindelliana, S. histata, S. elegans, S. dephnoides,* and *S. viminalis* are introduced varieties (Vishnu-Mittre 1966). Other common introduced elements are *Morus alba, M. nigra, Robinia pseudoacacia,* and *Biota orientalis*.

Over-grazing is another factor responsible for degradation of vegetation and soil. The carpet of grass that springs up is in no time browsed to the ground level, thus creating soil erosion. The decreased shoot cover and root growth expose the land to erosive forces such as the cutting effect of runoff water. The trampling by animals itself has a destructive effect on soil structure. The cattle are also left in the forests to browse whatever they can. Lopping of the leaves of trees provides additional fodder. Extensive lopping has resulted in the reduction of *Betula utilis* and near extinction of *Quercus* species in the valley. The impact of early man and his grazing animals upon vegetation in Kashmir has also been inferred from palynological studies (Dodia et al. 1985). The selective use of fir at Beba Rishi and Yus Maidan (Sharma and Vishnu-Mittre 1968), the extinction of oaks and alders in the Kashmir valley, and the poor values of grasses in Kashmir pollen diagrams are attributed to intensive activity of grazing animals (Vishnu-Mittre 1966, Vishnu-Mittre and Sharma 1966).

CHAPTER 9

ORIGIN AND HISTORY OF AGRICULTURE

The archaeobotanical investigations coupled with palynological investigations can also be employed in depicting the origin and history of agriculture. Man's struggle for comfortable existence and evolution from food gathering to food producing is reflected in his effort to grapple with his environment at every level. When man began to practise agriculture, he had to clear land for farming and had to disturb the natural vegetation. The events of the earliest cultivation by man could be depicted through pollen analysis of lake and swamp deposits. Such studies in Kashmir at Haigam lake and Anchar lake trace the beginnings of agriculture in the valley to 4000 years B.P. and at about the same time undoubted evidence of agriculture is available from archaeological excavations (Vishnu-Mittre and Sharma 1966, Dodia et al. 1985). The earliest evidence of agriculture in India goes back to 7000–8000 years B.P., in the pollen analysis of the salt lakes of Rajasthan (Singh 1971), the evidence of rice cultivation at Koldihawa, and barley cultivation at Mahagara (Sharma and Mandal 1980, Sen Gupta 1985). According to this evidence, Kashmir falls some 3,000 to 4,000 years behind in cultivation practices. But is it really true? Indeed, more excavations and pollen analyses are needed before arriving at a sound conclusion.

The first report on the plant remains from Kashmir archaeological excavations came when Vishnu-Mittre (1966) identified seeds of weed plants like *Lithospermum arvense, Medicago* spp., *Lotus corniculata*, and *Ipomoea* spp. from Neolithic levels at Burzahom and commented that inhabitants were essentially food gatherers. Our studies on the site have revealed the cultivation of wheat, barley, lentil, and pea from the Neolithic phase. This is further confirmed by the retrieval of wheat and barley at Gofkral (Sharma 1982) and wheat, barley, and rice at Semthan at a slightly later stage as evidenced by the present study.

The evidence from these three sites in the valley, which cover a time span of 4325 to 1000 years B.P., gives a clear picture of the origin and progressive development of agriculture. The earliest cultivation in Kashmir, as the present evidence suggests, was of wheat, barley, lentil, and pea, which interestingly were the first crops to have been domesticated in the Near East, where agriculture first developed (Renfrew 1973). Subsequently, rice and species of *Phaseolus* were introduced. The agriculture was further augmented by the adoption of horticultural fruits like *Juglans regia, Prunus armeniaca, P. amygdalus,* and *P. persica*.

Taking the cropping pattern and progressive development of agriculture into consideration, a significant evolutionary trend is indicated as follows:

1. *Single cropping*: During the third millennium B.C. only the rabi or winter crops were cultivated. This is indicated by the grains of exclusively winter crops such as wheat, barley, lentil, and pea at Burzahom period I and Gofkral period I (Sharma 1982).

2. *Double cropping*: During the second millennium B.C. two crops a year were cultivated: winter crops and kharif or summer crops such as rice and *Phaseolus* spp. as evidenced during the close of Neolithic period II at Gofkral and period I at Semthan.

3. *Mixed cropping*: Toward the close of the second millennium B.C. and during the first millennium B.C., double cropping was augmented by the cultivation of horticultural fruits like *Prunus* spp. and *Juglans* spp., as well as some trees for fodder and fuel, like *Salix, Populus, Celtis Ulmus,* and *Fraxinus*. This was later supplemented by the introduction of more useful trees like *Morus alba*, which marks the beginning of sericulture industry, and *Platanus orientalis*.

So based on this evidence the history of agriculture in the valley can be depicted as follows:

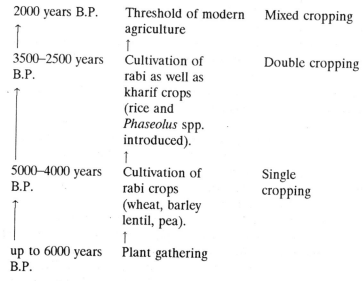

CHAPTER 10

STATE OF ECONOMY

In this chapter we shall consider the plant economy under different age groups collectively so as to point out in which particular aspect progress is evident and in which there are signs of retrogression.

10.1 AGRICULTURE

10.1.1. Burzahom (fig. 10.1)

PERIOD I: NEOLITHIC I (2375–1700 B.C.)

The remains of agricultural crops recovered from the Neolithic I period of occupation at Burzahom are extremely interesting. These belong to cereals, a pulse, and some horticultural crops. Some weed seeds have also been recovered. These plant remains were found neither in heaps nor scattered on the floor but embedded in occupation soil and mud clods. How could this have been brought about? Evidently whatever was produced was consumed and every grain was taken care of.

The presence of grains, seeds, and fruits in the mud clods still remains unaccounted for. Maybe a few grains got dislodged and thrown about during the process of threshing. These ultimately found their way into the crevices of the floor of the dwelling pits, where frequent treading by inhabitants forced the grains deep into the floor to be retrieved during excavation. The fact that the weeds and endocarps of stone fruits were obtained from the same mud clods that revealed food grains indicates that these have been deliberately collected by the inhabitants.

Cereals constitute 73.68% of the total agricultural economy of this period. Wheats (*Triticum aestivum*) and *T. sphaerococcum*) constitute 78.5% of the cereals and barley (*Hordeum vulgare*) 21.5%. The pulse recovered belongs to *Lens culinaris*, constituting 1.75% of the economy. The horticultural crops make up 10.5% of the agricultural economy, being represented

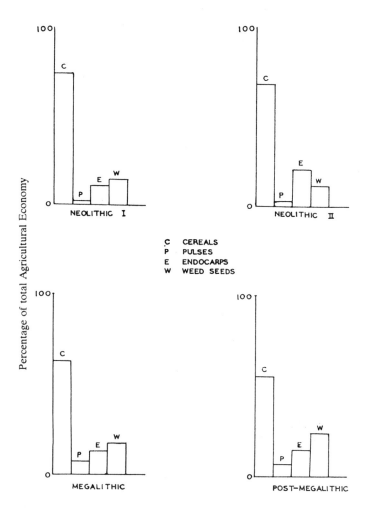

Figure 10.1: Period-wise plant economy at Burzahom.

by peach (*Prunus persica*), apricot (*P. armeniaca*), and walnut (*Juglans regia*), which constitute 33.3%, 16.6%, and 50% respectively of the horticultural crops. The associated weeds constitute 14% of the economy and comprise *Medicago* spp., *Vicia* spp., *Galium aparine*, *Astragalus* spp., and *Ipomoea* spp.

Both wheat and barley are well-known winter crops. Their water requirement is moderate and they do well under irrigation. Present experience shows that these crops do fairly well even in the absence of a proper irrigation system. The inhabitants of Burzahom during the Neolithic I period were growing winter crops only and practised a single cropping system of agriculture.

The pulse *Lens culinaris* is also a cold weather crop (Kachroo and Arif 1970) and can be grown on unirrigated soils. This further confirms that only winter crops were grown.

Recovery of fruit stones belonging to *Prunus persica, P. armeniaca*, and *Juglans regia* indicate that the inhabitants of the first period at Burzahom were aware of the edible quality of these horticultural crops.

PERIOD II: NEOLITHIC II (1700–1000 B.C.)

The plant assemblage of this phase consists of cereals, a pulse, horticultural crops, and associated weeds. Cereals constitute 67.8% of the total economy and comprise *Triticum aestivum, T. sphaerococcum*, and *Hordeum vulgare* as in period I. Wheats constitute 73.68% of the cereals and barley 26.32%. The pulse recovered belongs to *Lens culinaris*, constituting 1.9% of the economy. Horticultural crops (endocarps) constitute 19.6% of the economy and comprise *Prunus amygdalus, P. armeniaca*, and *Juglans regia*, which constitute 39.5%, 21%, 39.5% and 9% respectively of horticultural crops. Associated weed seeds recovered constitute 10.7% of the economy and belong to *Melilotus albus, Galium tricorne, Lithospermum arvense*, and *Ipomoea* spp. One seed belonging to *Vitis vinifera* has also been recovered from this period.

The cereals and pulses continue to be the same as in period I. One addition to the plant economy is *Prunus amygdalus*. *Melilotus albus, Galium tricorne*, and *Lithospermum arvense* are the new taxa among the weed seeds.

The overall economy suggests that there is certainly some advancement over period I. The most important addition is almond, *Prunus amygdalus*, for which the valley is famous even today. The presence of so many weed seeds also indicates high agricultural activity.

PERIOD III: MEGALITHIC (1000–600 B.C.)

Agricultural economy consists of cereals, pulses, horticultural crops, and weed seeds. Cereals constitute 62.8% of the economy and belong to wheats (*Triticum aestivum* and *T. sphaerococcum*), barley (*Hordeum vulgare*), and rice (*Oryza sativa*). Wheats constitute 65.9%, barley 25%, and rice 9.1% of the cereals.

This period is characterized by the introduction of rice, a totally new crop with different requirements for its growth. It is a summer crop requiring plenty of water for optimal growth. The adoption of rice indicates that the inhabitants had started raising two crops a year, rice in summer and wheat or barley in winter. The introduction of rice definitely indicates a change in food habits. Rice continues to be the main crop of the valley at present.

Pulses of this phase constitute 7.2% of the economy and belong to

Lens culinaris and *Pisum sativum*. Pea, *Pisum sativum*, is the new addition. The pea requires a cool growing season, abundant rainfall, and a relatively humid climate (Martin and Leonard 1967, Purseglove 1977).

Horticultural crops constitute 12.8% of the economy, comprising *Prunus persica* (68.6%), *P. domestica* (15.7%), and *Juglans regia* (15.7%). *Prunus domestica* is the new addition to horticulture.

Weed seeds continue to be the same as in periods I and II, constituting 17.2% of the economy and being represented by *Vicia* spp., *Astragalus* spp., *Melilotus albus, Galium aparine,* and *Lithospermum arvense.*

Period III is important because it is the beginning of rice cultivation at Burzahom, marking the shift from single cropping to double cropping.

PERIOD IV: POST-MEGALITHIC (600 B.C.–200 A.D.)

Cereals, pulses, horticultural crops, and weed seeds constitute the agricultural economy of this phase also. Cereals constitute 54.4% of the economy, of which wheats (*Triticum aestivum* and *T. sphaerococcum*) constitute 60%, barley (*Hordeum vulgare*) 25.4% and rice (*Oryza sativa*) 14.5%. Pulses constitute 6.9% of the economy, of which *Lens culinaris* accounts for 71.4% and *Pisum sativum* 28.6%. Endocarps constitute 14.8%, represented by *Prunus persica* (62.5%), *P. domestica* (12.5%), and *P. amygdalus* (25.0%). Weed seeds recovered make up 23.7% of the economy and belong to *Medicago* spp., *Astragalus* spp., *Galium aparine, G. tricorne, Lithospermum arvense, Vicia* spp., and *Ipomoea* spp.

The important feature of this phase has been the recovery of weed seeds in large numbers with a great variability which indicates agricultural intensification.

The overall evidence indicates that the agricultural economy of this period has been more or less a continuation of what was achieved in period III.

10.1.2. Semthan (fig. 10.2)

PERIOD I: PRE-N.B.P. (3500–2600 YEARS B.P.)

The plant assemblage recovered from this phase is quite interesting. Just above the natural soil, rice (*Oryza sativa*), barley (*Hordeum vulgare*), and wheat (*Triticum aestivum* and *T. sphaerococcum*) are recovered. Associated with these cereals are the pulses lentil (*Lens culinaris*) and green gram (*Phaseolus aureus*) and endocarps of *Juglans regia, Prunus armeniaca,* and *Celtis australis.*

Cereals constitute about 85% of the total seeds recovered from this phase. Of these rice constitutes 55.8%, wheat 30.8%, and barley 14.4%. Rice is a summer crop and needs plenty of water for optimal growth. Wheat and barley are well-known winter crops and require moderate irrigation. This indicates that the inhabitants raised two crops a year, rice in the summer and wheat or barley in the winter, which indicates that the agriculture was very advanced and the people who had settled at Semthan carried with them a good knowledge of agriculture and were well acquainted with the double cropping system. It is known through Vedic literature that 'two harvests a year were gathered' (Bose et al. 1971) and the Vedic period according to some scholars begins about 1500 B.C. (Kosambi 1965, Thapar 1966, Sankalia 1971). Further, the practice of growing two crops a year was in vogue much earlier than the Vedic period, as revealed by the retrieval of rice and barley together at Atranjikhera Phase I, 2000–1500 B.C. (Chowdhury et al. 1977). However, we are still unaware of the cropping pattern and rotation practices of the Harappans although a meagre evidence of mixed cropping is revealed by the Kalibangan furrowed field (Vishnu-Mittre and Savithri 1982).

As regards pulses, they constitute just 4% to 5% of the total seed economy of the phase and are represented by *Lens culinaris* and *Phaseolus aureus*. *Lens culinaris* is a cold weather unirrigated crop and as a pulse ranks inferior to green gram (Kachroo and Arif 1970), whereas *Phaseolus aureus* is a crop that can be grown in spring or summer and needs a well-distributed, moderate rainfall. It can be grown on unirrigated lands and is a much valued pulse (Anonymous 1969b, Kachroo and Arif 1970). The recovery of a summer crop and a winter crop among the pulses as well further confirms that two crops a year were harvested.

The endocarps recovered constitute 9.8% of the seed economy and are represented by *Prunus armeniaca, Juglans regia,* and *Celtis australis*. Recovery of *Juglans regia* and *Prunus armeniaca* hints at the probability of horticultural crops being grown or collected by the inhabitants, which further marks their advancement in agriculture. The recovery of endocarps of *Celtis australis* is a bit intriguing. However, a point may be mentioned here that among the wood remains recovered from this phase there was a piece belonging to *Celtis* wood. It is quite possible that a few dry fruits attached to the tree, which was felled, somehow found their way into the mud clods.

The whole evidence leads to the conclusion that the settlers at Semthan had a good knowledge of agriculture and were utilizing cereals, pulses, and some fruits as well.

PERIOD II: N.B.P. (2600–2200 B.P.)

The plant assemblage in this phase consists of cereals, pulses, endocarps, and weed seeds. Of these, cereals constitute 74.8% of the total economy and are represented by the three cereals recovered from the pre-N.B.P. phase, i.e., *Oryza sativa, Triticum aestivum,* and *Hordeum vulgare,* with the addition of few caryopses of *Avena* spp. (comparable to *A. fatua*). None of the grains

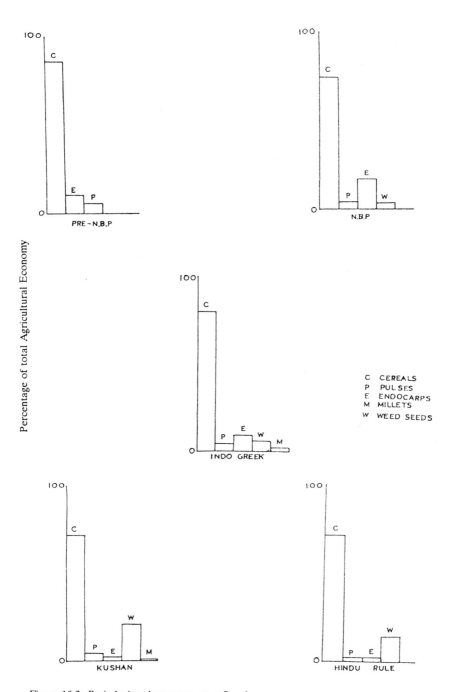

Figure 10.2: Period-wise plant economy at Semthan.

referable to *Triticum sphaerococcum* could be recovered. Among the cereals, rice constituted 17.5%, wheat 25%, barley 56.25%, and oats 1.17%.

Pulses constitute 4.38% of the economy and show a lot of variety, being represented by *Phaseolus mungo* (26.6% of the pulses), *P. aureus* (33.3%), *P. aconitifolius* (6.6%), *Lens culinaris* (6.6%), and *Pisum sativum* (6.6%). All these belong to the family of legumes, which are well known for their ability to fix atmospheric nitrogen and thereby improve the fertility of soil on which they grow. It may be pointed out that Vedic literature has emphasized the rotation of crops in order to maintain soil fertility and crop productivity (Bose et al. 1971). The ancient Kashmiris, too, were aware of this fact. New additions to the Pre-N.B.P. pulses are *Phaseolus mungo*, *P. aconitifolius*, and *Pisum sativum*. *Phaseolus mungo* is a highly prized pulse and is grown in the valley as a summer (kharif) crop. *Phaseolus aconitifolius* is the most drought resistant of the kharif pulses (Anonymous 1969b). *Pisum sativum* requires a cool growing season and flourishes best with a fairly abundant rainfall (Martin and Leonard 1967) and a relatively humid climate (Purseglove 1977). Such a large variety of pulses indicates that the inhabitants were well aware of the utility of pulses.

The endocarps recovered constitute 17.25% of the total economy of which *Prunus cerasus* accounts for 61.2%, *P. armeniaca* 15.2%, *Juglans regia* 16.9%, and *Celtis australis* 6.7%. *Prunus cerasus* is the new addition to Pre-N.B.P. endocarps. It is a wild edible fruit tree which might have been collected by the inhabitants from the nearby forests.

Weed seeds constitute 3.5% of the plant economy and are represented by *Lithospermum arvense* and *Vicia/Lathyrus* spp. Both are the common weeds of the local field crops, particularly the rainfed crops.

From these facts one is led to presume that there is certainly a great advancement in the agricultural economy over period I.

PERIOD III: INDO-GREEK (2200–2000 B.P.)

The agricultural economy of this phase is made up of cereals, a millet, pulses, endocarps, and weed seeds. The cereals constitute 80% of the economy and the interesting feature is the addition of oat grains that could be referred to *Avena fatua* as well as *Avena sativa*. Rice (*Oryza sativa*) constitutes 23.5%, wheat (*Triticum aestivum*) 4.4%, barley (*Hordeum vulgare*) 67.6% and oats (*Avena fatua* and *A. sativa*) 4.4% of the cereal economy. *Avena* is essentially a winter crop and can also be grown as a summer crop. Perhaps the utility of this crop as a fodder had become known to the early farmers who had begun its cultivation.

Recovery of a few grains of *Panicum* spp. is quite interesting and significant too. *Panicum* is among the unconventional cereals or 'poor man's food' to which man resorted in times of scarcity or famine (Vishnu-Mittre

1985). It is also cultivated and grown exclusively on unirrigated poor soils (Anonymous 1969b). Was there any kind of famine or scarcity during this period? Literary records in this regard are lacking.

Pulses constitute 4.7% of the economy and are represented by seeds of *Phaseolus mungo* and *Phaseolus aureus*. Among the endocarps which constitute 9.4% of the economy, *Prunus persica* is the new addition. Others recovered are *Prunus armeniaca* and *Juglans regia*. Weed seeds contitute 5.6% of the economy and are represented by *Lithospermum arvense, Vicia/Lathyrus* spp., and an unidentified leguminous seed.

The overall evidence reveals that during period III, that is, toward the beginning of the Christian era, quite a large number of plant species had been cultivated, indicating a continuous upward trend in the rural economic and agricultural development.

PERIOD IV: KUSHAN (2000–1500 B.P.)

Cereals constitute 72.3% of the total seed economy of this phase. Of these, rice constitutes 28%, wheat 39.3%, barley 27.8%, and oats 4.5%. Wheat is represented by two species, *Triticum aestivum* and *T. sphaerococcum*.

An interesting feature of this phase is the recovery of a few seeds of *Setaria* spp. It is an important millet and could have been adopted as an unconventional cereal.

Pulses constitute 4.3% of the economy and are represented by *Pisum sativum, Lens culinaris, Phaseolus aureus, P. mungo*, and *P. aconitifolius*, as in the N.B.P. phase. Endocarps, which constitute 1.8% of the economy, are represented by *Prunus armeniaca, P. persica*, and *Juglans regia*.

Another interesting feature of this phase has been the recovery of a large number of weed seeds, which constitute 21.5% of the total seed economy and are represented by *Galium aparine, G. asperuloides, G. tricorne, Lithospermum arvense, Vicia/Lathyrus* spp., *Medicago* spp., and some unidentified seeds. The presence of various types of weed seeds indicates a progressive trend in the rural village farming activity. It is also deducible that some of the weeds were deliberately collected for feeding cattle, which in turn, indicates an advanced animal husbandry.

From the above evidence, it becomes clear that rural and agricultural economy was approaching a climax during this period. A lot of cereals, pulses, and fruit trees were being cultivated. The ancient literature, accounts of travellers, and archaeological discoveries describe that the period between the fifth and ninth centuries is the period of maximum prosperity in the valley (Ray 1957). The Kushan period dates up to the threshold of this prosperous period in the valley.

PERIOD IV: HINDU RULE (1500–1000 B.P.)

The agricultural economy is represented by cereals, pulses, endocarps, and weed seeds. Cereals constitute 71.7%, of which rice (*Oryza sativa*) constitutes 32.3%, wheat (*Triticum aestivum*) 23.9%, barley (*Hordeum vulgare*) 38%, and oat (*Avena sativa*) 5.63%. Pulses constitute 2.02% and are equally represented by lentil and pea. Endocarps constitute 12.12% and are represented by *Prunus armeniaca*, *P. persica*, and *Celtis australis*. Weed seeds constitute 14.14%, being represented by *Galium tricorne*, *Lithospermum arvense*, *Vicia/Lathyrus* spp., *Melilotus albus*, and an unidentified seed.

These evidences lead to the conclusion that the rural economy during this period was more or less a continuation of what was established in the Kushan period, 2000–1500 B.P. There is no new addition to the plant economy.

10.2. FORESTRY

10.2.1. Burzahom

PERIOD I: NEOLITHIC I

At the outset it seems pertinent to point out that small pieces of charred wood that were recovered came from the same source as cereals, pulses, endocarps, and weed seeds. It is not quite clear how these got into the floor of the house; however, these wood remains are significant. In this context Braidwood and Howe (1960) have aptly said, "We were very conscious of the fact that plants and animals do not domesticate themselves, nor does an environment domesticate them. The domesticator is man."

The wood remains recovered from this period belong to *Pinus wallichiana*, *Betula utilis*, *Quercus* spp., *Ulmus* spp., *Populus* spp., *Aesculus indica*, *Crataegus oxyacantha*, and *Ficus* spp. Of these, *Ficus* spp. is an introduced element. The others must have been available in the near vicinity. That these well-valued timbers were known to the inhabitants indicates their knowledge about the utility of available forest products.

PERIOD II: NEOLITHIC II

The woods identified from this phase belong to *Picea smithiana*, *Quercus* spp., *Fraxinus excelsior*, *Vitis vinifera*, *Populus* spp., *Aesculus indica*, and some leguminous wood. The new taxa include *Picea smithiana*, *Fraxinus excelsior*, *Vitis vinifera* and some leguminous wood. More and more woods were being used by the ancient inhabitants, indicating advancement in the use of forest wealth over period I.

PERIOD III: MEGALITHIC

Pinus wallichiana, Platanus orientalis, Populus spp., *Aesculus indica, Buxus* spp., and some leguminous woods were identified from this phase. *Buxus* spp. and *Platanus orientalis* appear for the first time. The present and past distribution of these plants in the valley indicate that both the plants have been obtained from outside.

PERIOD IV: POST-MEGALITHIC

The wood remains of this phase belong to *Cedrus deodara, Juglans* spp., *Celtis australis, Ulmus wallichiana, Fraxinus excelsior,* and *Aesculus indica*. The two important timbers used for the first time during this phase are *Cedrus deodara* and *Celtis australis*.

The ancient inhabitants have been trying and using more and more woods, adopting some while discarding others.

10.2.2 Semthan

PERIOD I: PRE-N.B.P.

The wood remains recovered from this phase belong to *Pinus wallichiana, Picea smithiana, Cedrus deodara, Juglans* spp., *Celtis australis, Quercus* spp., and some leguminous wood. The use of these few woods among the large number available to the inhabitants is very significant. Considering the economic stage of their existence, it is remarkable that these well-valued timbers were known to them in preference to many others that formed the virgin forests around them.

PERIOD II: N.B.P.

The wood remains recovered from this phase belong to *Pinus wallichiana, Abies pindrow, Cedrus deodara, Quercus* spp., *Ulmus wallichiana, Fraxinus excelsior, Prunus* spp., *Betula utilis,* and *Ficus* spp. Addition of so many timbers is a clear indication of the advancement in use of forest wealth over period I.

PERIOD III: INDO-GREEK

Wood remains recovered from this phase belong to *Picea smithiana, Salix* spp., *Aesculus indica, Viburnum* spp., and *Ulmus* spp. *Aesculus, Salix,* and *Viburnum* are the new woods that were tried and used by the inhabitants. It is quite evident that people were trying more and more woods in order to know their utility and economic potential, which is an indication of economic advancement.

PERIOD IV: KUSHAN

The timbers recovered from this phase belong to *Cupressus* spp., *Juglans* spp., *Populus* spp., *Parrotiopsis jacquemontiana, Crataegus oxyacantha, Celtis australis, Aesculus indica, Pyrus pashia, Betula utilis,* and *Fraxinus excelsior.* The new woods used during this period are *Cupressus* spp., *Parrotiopsis jacquemontiana, Crataegus oxyacantha, Pyrus pashia,* and *Prunus* spp. Thus, if the agricultural economy of this phase depicts the climax of advancement, the forestry economy is also quite illuminating.

PERIOD V: HINDU RULE

Wood remains belonging to *Cupressus* spp., *Acer caesium, Platanus orientalis, Morus alba, Parrotiopsis jacquemontiana, Crataegus oxyacantha,* and *Prunus* spp. have been recovered from this phase. *Platanus* and *Morus* are newly introduced. *Acer caesium* has also been used for the first time.

The overall evidence on ancient forestry indicates that the relationship between the woody trees and man in Kashmir is of great antiquity and with the passage of time wood has been put to innumerable and varied uses by prehistoric man. The same woods are not repeated in all the periods. With time the inhabitants gradually gained experience of the quality of different woods and consequently new woods were tried and used while others were omitted. If in our advanced stage of technology we make use of different characteristics such as strength, workability, durability, and density in our selection of woods for a vast range of primary and secondary products, prehistoric man, too, was aware of the best wood for burning, for warmth, for building, and for making tools and other items.

From the domestication point of view it may be emphasized that man discovered fire long before he started cultivating food plants. For fire, the wood was a prerequisite. Thus the kinship between wood and the early man is very antique. However, evidence for the use of wood is available in the late Palaeolithic Age in England and Germany, where entire spears made of wood have found (Hawkes and Woolley 1963). Scanning through the information available in India, one finds that fire as a tool was known to early man almost at the commencement of the Holocene period (8000 B.C.) at Sarai Nahar Rai, where the charred bones of animals provide a testimony to the deliberate use of fire for cooking purposes (Sankalia 1974). Coupled with this is the evidence of several hearths, suggesting incipient community life. Pollen studies, too, provide evidence of fire as early as that but it remains to be determined whether it was a natural or man-made fire (Vishnu-Mittre 1974b).

Initially man might have used any kind of wood whatsoever available to him. As he got experienced, he began to use preferred woods and discard others. This resulted in the transportation of various woods from far off places, which further resulted in contact with people living in distant localities and provided links for exchange of ideas. In the course of time, evidently, man preferred those woody plants which besides providing wood of good quality also provided edible fruits, leaf, vegetables, or fodder. Here again, selection played an important role. If such trees were not available locally, man must have brought them from far off places and in the course of time tried to grow them in his own locality, a process which still continues. This hypothesis is substantiated by the evidence presented above, but more work is needed on the early prehistoric sites to delineate the process at regional as well as global levels.

CHAPTER 11

STATISTICS IN PALAEOETHNOBOTANY

Over the past few years, palaeoethnobotanists have made an increasing effort to apply their data to the questions of cultural process: the evolution of early nutritional systems, cultivation strategies, long-term stability of subsistence strategies, and the process of agricultural intensification, among other current research problems in archaeology. Ford (1979) details the evolution of this trend and the emergence of anthropological ethnobotany. Several recent publications (Asch and Asch 1975, Asch et al. 1979, Dennel 1972, Minnis 1981, Pearsall 1983) have emphasized the need for caution in the interpretation of archaeological seed assemblages. The lack of direct correlation between raw seed counts or percentages and dietary importance of the plant is well understood by most ethnobotanists who routinely include in their analyses caution about the bias produced by differential preservation of botanical materials archaeologically. A variety of quantitative approaches have been applied to circumvent this problem (Pearsall 1983). These include chi-square analysis, intensity of occupation, species diversity, and standard scores.

Accordingly, we tried to subject our data (table 11.6) to statistical analysis. It was realized that the data from Semthan is more suitable for such an analysis. Hence we restrict our statistical evaluation to the evidence from Semthan. In addition to the parameters applied by early workers, we have explored the possibility of application of some other parameters. It has been found that species evenness, species richness, and more so indices of similarity also prove quite useful.

For the statistical analyses, data from an equal number of flotation samples of equal volume from each phase was used.

Measures employed in this analysis were calculated as follows:

Chi-square values were determined using the following formula:

$$X^2 = \frac{(X_{Ob} - X_{Ex})^2}{X_{Ex}}$$

where X_{Ob} = observed number of seeds/charcoals of a taxon.
X_{Ex} = expected number of seeds/charcoals of the taxon.

Intensity of occupation was calculated as the total count of charred wood summed by phase.

For species diversity, Shannon and Weaver (1949) information index for finite population (H) was calculated in each phase as

$$H = -\Sigma \left(\frac{N_j}{N}\right) \ln \left(\frac{N_j}{N}\right)$$

where N_j = total number of charcoals of a taxon in the phase.
N = total number of charcoals in the phase.
ln = natural log.

Species richness was measured after Margalef (1957) as

$$r = (S - 1)/\ln N$$

and species evenness after Shannon and Weaver (1949) as

$$e = H/\ln S$$

where S = total number of species present in a phase.
N = Total number of charcoals in the phase.
H = Species diversity.

Standard scores or standard deviation units (Z) were calculated after Blalock (1972) as

$$Z = \frac{X - \bar{X}}{s}$$

where $\bar{X}$ = Occurrence of a seed taxon in a phase.
X = Mean occurrence of taxon in all the phases.
s = Standard deviation of a taxon for the mean.

Index of similarity (S) between pairs of phases has been computed after Wolda (1981) as

$$S = \frac{2C}{A + B}$$

where A and B are the total species recorded at each of a given pair of phases and C the total number of species recorded at both the phases.

Index of similarity (T) has been computed after Thibodeau and Nickerson (1985) as

$$T = \Sigma \frac{(N_j a + N_j b) - (N_j a - N_j b)}{(Na + Nb)}$$

where a and b are the subscripts labelling two phases between which similarity is to be determined; all other notations are as described above.

11.1. BEHAVIOUR OF INDIVIDUAL PLANT GROUPS

The behaviour of cereals, pulses, endocarps, and weed seeds from periods I to V is shown in fig. 11.1. It is seen that the cereals continue to dominate over other plant groups and represent 70% to 80% of the total economy of each period. Pulses maintain a low frequency of 2% to 5% in all the phases. The endocarps show maximum frequency of 17.25% in period II and a very low frequency of 1.79% in period IV. The weed seeds are absent from period I but show an increasing frequency thereafter, showing maximum frequency of 21.58% in period IV.

Among the cereals (fig. 11.2), rice shows a frequency of 55.8% in period I, then declines to 17.5% in period II, and thereafter shows a continuous increase. It has already been suggested that rice was introduced during 1500–1000 B.C., at the time of period I at Semthan. Its very high frequency at this time is quite surprising as wheat and barley were established crops at that time which show frequencies of 30.8% and 14.4% respectively. After period II, rice shows a steadily increasing frequency. Further, it is seen that wheat declines to its minimum frequency of 4.4% in period III and then increases to 39% in period IV. On the other hand, barley, after having lower values in period II, goes up substantially, reaching a value of about 67% in period III, and thereafter shows a decline. Oats, being introduced during period II, maintain a low frequency throughout but show a continuous slight increase.

It seems reasonable to conclude that the settlers at Semthan started with the cultivation of wheat, barley, and rice, barley and wheat initially being preferred to rice. Barley was much preferred even up to a much later time as it constitutes about two-thirds of the total economy of period III. However, rice continues to replace wheat and barley toward the later periods.

11.2. CHI-SQUARE ANALYSIS

Chi-square analysis of seed and charcoal data has been performed. The data matrix for cereals, pulses, endocarps, weed seeds, and charcoals is presented in tables 11.1–11.5.

Chi-square analysis of cereals revealed a chi-square value of 82.6 (df = 30), which indicates a high significant difference ($p<0.01$). Thus the cereal taxa do not appear to be randomly distributed among time periods. Periods I and III account for most of the lack of randomness between periods. Among the individual cereals, *Avena fatua* and *A. sativa* show the highest randomness in distribution.

For pulses a chi-square value of 15.1 (df = 25) indicates a significant difference of 0.90, revealing that these taxa are randomly distributed in time

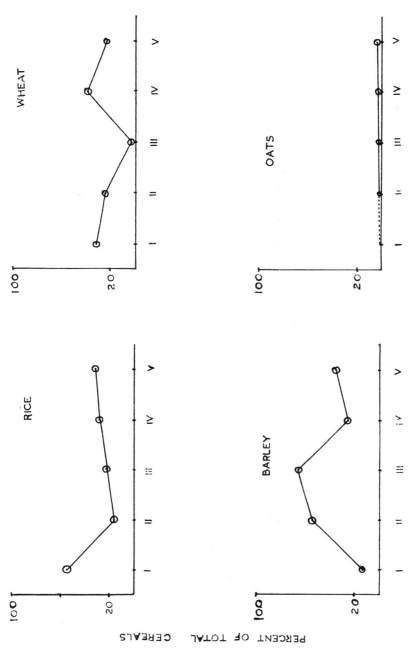

Figure 11.1: Behaviour of major agricultural plant groups over time.

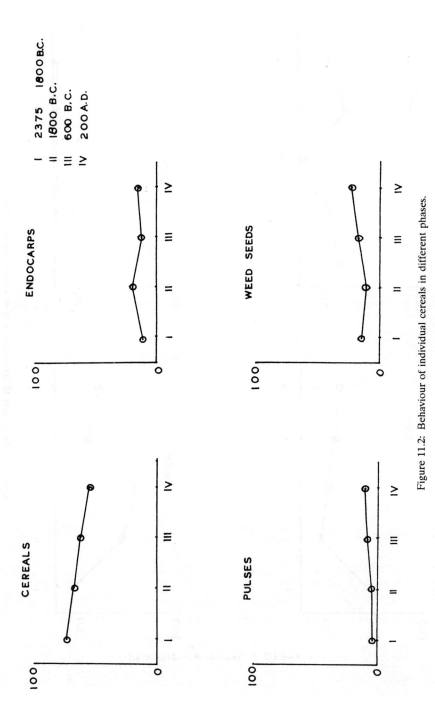

Figure 11.2: Behaviour of individual cereals in different phases.

periods. However, this very high randomness can be due to very small counts in the time periods and unevenness of data.

The endocarps and weed seeds show chi-square values of 47.0 (df = 25) and 21.03 (df = 16), indicating significant differences of 0.01 and 0.20 respectively. These values indicate that weed seeds are more randomly distributed then endocarps among the time periods. Most of the lack of randomness is accounted for by *Lithospermum* among the weeds and *Prunus cerasus, P. persica*, and *Celtis australis* among the endocarps. *Juglans* among the endocarps and 'others' among the weeds appear to be most randomly distributed.

A chi-square value of 2018.1 (df = 96) for the wood charcoal data indicates a very high significant difference ($p \ll 0.001$). Therefore, wood taxa do not appear to be randomly distributed among the time periods. Chi-square analysis was also performed to determine if non-random distributions existed within wood type categories or within time periods. The analysis showed that period III accounted for most of the lack of randomness between periods, followed by periods V, II, I and IV. Among the taxa most of the lack of randomness is accounted for by *Pinus, Picea, Abies, Cedrus, Juglans, Ulmus, Salix, Viburnum, Acer, Celtis, Aesculus, Morus*, and *Populus*. This conclusion has been drawn by comparing the observed (Ob.) and expected (Ex.) values for the matrix points in the row or column where high chi-square values occurred. For example, much of the deviation from chance giving a high chi-square value for period III row (629.7) occurs at *Picea, Ulmus, Salix* and *Viburnum*, for period V row at *Morus* and *Acer*, and for period II at *Pinus, Abies* and *Cedrus* (compare observed and expected values for these taxa).

The data matrix for charcoals was also used to know the behaviour of gymnosperm and angiosperm woods collectively over time. The results show that during periods I and II, gymnosperms account for 31.6% and 39.9% respectively. Their percentage declines to 22.1% in period III, 11.2% in period IV, and 17.5% in period V, with corresponding increases in the angiosperm woods. This can be attributed to the availability of gymnosperms in the near vicinity during the earlier periods. As a result of large-scale exploitation, these were restricted to the higher mountains leading to utilization and adoption of more and more broad-leaved elements, which could also have been easily grown by the inhabitants. This presents an example of maintenance of a traditional fuel system. The overall evidence suggests that selection played a very important role in the adoption and domestication of various plant species which have not been randomly chosen.

Table 11.1. Chi-square analysis of cereals and time periods

Period		O.s.	T.a.	T.s.	H.v.	A.f.	A.s.	Row total	Row chi	Sign. level
I.	Ob.	22	8	4	5	0	0	39	29.8	0.001
	Ex.	11.32	8.76	0.93	16.57	0.93	0.51			
II	Ob.	26	28	0	64	3	0	121	9.9	0.10
	Ex.	35.13	27.14	2.87	51.40	2.87	1.59			
III	Ob.	16	3	0	46	2	1	68	22.4	0.001
	Ex.	19.74	15.25	1.61	28.89	1.61	0.90			
IV	Ob.	29	33	5	28	4	2	101	14.2	0.02
	Ex.	29.32	22.65	2.39	42.39	1.34	1.34			
V.	Ob.	17	13	0.	18	0	2	50	6.3	0.30
	Ex.	14.51	11.21	1.19	2.24	1.19	0.66			
Col.	Total	110	85	9	161	9	5	379	82.6	
Col.	chi	13.8	15.2	18.7	27.1	3.5	4.3	82.6		
Sign.	lev	0.02	0.01	0.001	0.0001	0.70	0.50			

Overall chi square = 8.26 with 30 degrees of freedom significant at less than 0.01 level.
Taxa abbreviations: *Oryza sativa* (O.s.), *Triticum aestivum* (T.a.), *T. sphaecococcum* (T.s.) *Hordeum vulgare* (H.v.), *Avena fatua* (A.f.), and *A. sativa* (A.s.).
Ob.: observed.
Ex.: expected.

Table 11.2. Chi-square analysis of pulses and time periods

Period		P.a.	P.m.	P.ac.	L.c.	P.s.	Row total	Row chi	Sign. level
I	Ob.	1	0	0	1	0	2	4.0	0.50
	Ex.	0.50	0.50	0.17	0.25	0.58			
II	Ob.	2	2	1	0	2	7	1.25	0.95
	Ex.	1.75	1.75	0.58	0.88	2.04			
III	Ob.	2	2	0	0	0	4	4.0	0.50
	Ex.	1.00	1.00	0.33	0.5	1.17			
IV	Ob.	0	2	1	1	2	6	2.3	0.80
	Ex.	1.50	1.50	0.50	0.75	1.75			
V	Ob.	1	0	0	1	3	5	3.55	0.70
	Ex.	1.25	1.25	0.40	0.63	1.46			
Col. total		6	6	2	7	3	24	15.1	
Col. chi		3.1	3.0	1.7	3.9	3.4	15.1		
Sign. level		0.70	0.70	0.90	0.50	0.70			

Overall chi square = 15.1, with 25 degrees of freedom, significant at 0.90 level.
Taxa abbreviations: *Phaseolus aureus* (P.a.), *P. mumgo* (P.m.), *P. aconitifolius* (P.ac.), *Lens culinaris* (L.c.), and *Pisum sativum* (P.s.).
Ob: observed.
Ex: expected.

Table 11.3. Chi-square analysis of endocarps and time periods

Period		P.a.	P.p.	P.c.	J.r.	C.a.	Row total	Row chi	Sign. level
I	Ob.	1	0	0	1	3	5	14.1	0.01
	Ex.	1.21	0.43	1.81	1.03	0.52			
II	Ob.	5	0	21	7	2	35	10.7	0.05
	Ex.	8.45	3.02	12.67	7.24	3.62			
III	Ob.	3	2	0	3	0	8	7.9	0.20
	Ex.	1.93	0.69	2.90	1.65	0.83			
IV	Ob.	1	1	0	1	0	3	3.8	0.50
	Ex.	0.72	0.26	1.09	0.63	0.31			
V	Ob.	4	2	0	0	1	7	10.5	0.05
	Ex.	1.69	0.6	2.53	1.45	0.72			
Col.	total	14	5	21	12	6	58	47.0	
Col.	chi	5.3	11.3	13.8	2.8	13.8	47.0		
Sign.	level	0.30	0.05	0.02	0.8	0.02			

Overall chi square = 47.0, with 25 degrees of freedom significant at less than 0.01 level.
Texa abbreviations: *Prunus armeniaca* (P.a.), *P. persia* (P.p.), *P. cerasus* (P.c.), *Juglans regia* (J.r.), and *Celtis australis* (C.a.).
Ob.: observed.
Ex.: expected.

Statistics in Palaeoethnobotany

Table 11.4. Chi-square test of weed seeds and time periods

Period		Vicia	Lithospermum	Galium	Others	Row total	Row chi	Sign. level
I	Ob.	0	0	0	0	0	0	0
	Ex.	0	0	0	0	0		
II	Ob.	2	4	0	0	6	8.6	0.05
	Ex.	0.79	1.67	2.66	0.88			
III	Ob.	1	3	0	1	5	6.4	0.20
	Ex.	3.93	1.39	2.21	0.74			
IV	Ob.	5	6	23	6	40	4.0	0.50
	Ex.	5.25	11.14	17.71	5.90			
V	Ob.	0	4	4	2	10	2.1	0.70
	Ex.	1.31	2.79	4.43	1.47			
Col. total		8	17	27	9	61	12.1	
Col. chi		5.4	8.0	6.5	1.2	21.1		
Sign. level		0.20	0.05	0.20	0.90			

Overall chi square = 21.1, with 16 degrees of freedom, significant at 0.20 level.
Ob: observed.
Ex: expected.

Table 11.5. Chi-square analysis of wood taxa and time periods

Period		Pinus	Picea	Abies	Cedrus	Cupressus	Other Gymn.	Celtis	Juglans	Quercus	Fraxinus
		1	2	3	4	5	6	7	8	9	10
I	Ob.	31	18	0	22	0	28	29	38	15	0
	Ex.	10.33	7.16	4.11	7.42	3.84	35.77	9.41	11.26	3.71	7.42
II	Ob.	47	0	31	34	0	87	0	0	13	24
	Ex.	17.50	12.12	6.96	12.57	6.51	60.60	15.93	19.07	6.28	12.57
III	Ob.	0	36	0	0	0	37	0	0	0	0
	Ex.	11.60	8.03	4.61	8.33	4.31	40.15	10.56	12.64	4.17	8.33
IV	Ob.	0	0	0	0	16	64	42	47	0	32
	Ex.	25.17	17.42	10.00	18.07	9.36	87.12	22.91	27.43	9.04	18.07
V	Ov.	0	0	0	0	3	54	0	0	0	0
	Ex.	13.40	9.27	5.32	9.61	4.98	46.36	12.19	14.60	4.80	9.61
Col. Total		78	54	31	56	29	270	71	85	28	56
Col. Chi		141.3	152.6	107.1	101.2	32.3	20.8	95.4	123.8	59.6	46.5
Sign. Level		0.0001	0.0001	0.0001	0.0001	0.001	0.001	0.0001	0.0001	0.001	0.001

Statistics in Palaeoethnobotany

Period		Ulmus 11	Betula 12	Populus 12	Prunus 14	Aesculus 15	Salix 16	Viburnum 17	Parrotiopsis 18	Crataegus 19
I	Ob.	0	0	0	0	0	0	0	0	0
	Ex.	8.35	5.96	10.47	10.07	7.29	4.50	2.65	5.56	4.77
II	Ob.	26	21	19	23	0	0	0	0	0
	Ex.	14.14	10.10	17.73	17.06	12.34	7.63	4.49	9.43	8.08
III	Ob.	37	0	0	0	32	34	20	0	0
	Ex.	9.37	6.70	11.75	11.30	8.18	5.06	2.98	6.25	5.35
IV	Ob.	0	24	60	29	23	0	0	28	24
	Ex.	20.32	14.52	25.49	24.52	17.75	10.97	6.45	13.55	11.62
V	Ov.	0	0	24	0	0	0	14	12	0
	Ex.	10.82	7.72	13.56	13.05	9.44	5.84	3.43	7.21	6.18
Col.	Total	63	45	79	76	55	34	20	42	36
Col.	Chi	130.9	38.3	82.6	48.4	99.9	194.4	114.2	43.0	36.9
Sign.	Level	0.0001	0.001	0.001	.001	0.0001	0.0001	0.0001	0.001	0.001

(contd.)

Table 11.5. (Contd.)

		Pyrus	Acer	Platanus	Morus	Ficus	Other Ang.	Row total(1–25)	Row chi	Sign. level
		20	21	22	23	24	25			
I	Ob.	0	0	0	0	0	113	294	313.0	0.0001
	Ex.	2.25	2.78	1.59	2.38	1.46	123.48			
II	Ob.	0	0	0	0	11	162	498	378.2	0.0001
	Ex.	3.82	4.71	2.69	4.04	2.47	209.16			
III	Ob.	0	0	0	0	0	134	330	629.7	0.0001
	Ex.	2.53	3.12	1.79	2.68	1.64	138.60			
IV	Ob.	24	17	0	0	0	310	716	299.2	0.0001
	Ex.	5.49	6.78	3.87	5.81	3.55	300.73			
V	Ov.	12	0	12	18	0	213	381	398.0	0.0001
	Ex.	2.91	3.61	2.06	3.09	1.88	160.03			
Col.	Total	17	21	12	18	11	932	2219	2018.1	
Col.	Chi	35.6	101.2	57.9	86.8	37.9	29.5	2018.1		
Sign.	Level	0.001	0.001	0.001	0.0001	0.001	0.001			

Overall chi square = 2018.1, with 96 degrees of freedom, significant at less than 0.0001 level.
Ob.: observed.
Ex.: expected.

11.3. INTENSITY OF OCCUPATION

Intensity of occupation has been obtained from charred wood counts through time. The following results were obtained and are shown in fig. 11.3.

Period	Wood count
I: Pre-N.B.P.	294
II: N.B.P.	498
III: Indo-Greek	330
IV: Kushan	716
V: Hindu Rule	381

The total amount of charcoal resulting from deliberate burning can serve as a measure of the intensity of use of an occupation area (Asch and Asch 1975, Johannessen 1981a, 1981b). An occupation containing a much higher amount of charcoal under conditions of similar preservation and sampling is interpreted as having more intensive cooking or other hearth activity. Present studies at Semthan reveal that period IV (Kushan) shows the highest intensity of occupation (wood count = 716), followed by period II (N.B.P.; wood count = 498). The other three periods show some constancy in the intensity measure. It can be concluded that the intensity measure shows an increasing trend from periods I to IV with some decrease in activity in period III. In the last period the intensity measure declines, but remains slightly higher than in periods I and II.

11.4. SPECIES DIVERSITY (H), SPECIES RICHNESS (R) AND SPECIES EVENNESS (E)

For this analysis all the wood taxa and time periods were used. The following results were obtained and are depicted in figs. 11.4–11.6.

Period	Species diversity (h)	Species richness (r)	Species evenness (e)
I: Pre-N.B.P.	0.79	1.23	0.38
II: N.B.P.	0.92	1.77	0.37
III: Indo-Greek	0.71	1.03	0.36
IV: Kushan	0.87	1.82	0.34
V: Hindu Rule	0.67	1.35	0.30

Species diversity is a measure that takes into account both the total number of species or taxa present in a population and the abundance of each species (Pielou 1969). High diversity results when a large number of species are evenly distributed, that is, when it would be difficult to predict

230 *Palaeoethnobotany*

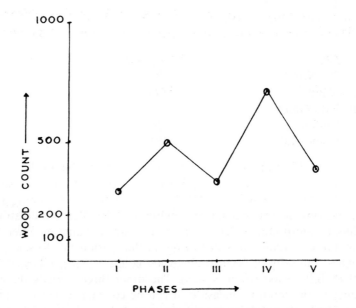

Figure 11.3: Intensity of occupation expressed as total wood count summed by phase.

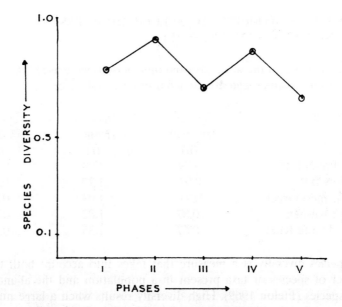

Figure 11.4: Species diversity of wood taxa expressed by phase.

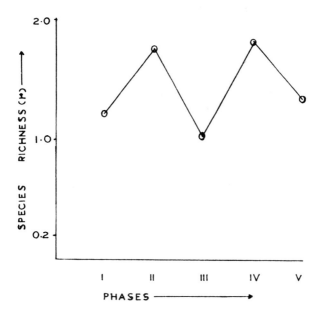

Figure 11.5: Species richness of wood taxa expressed by phase.

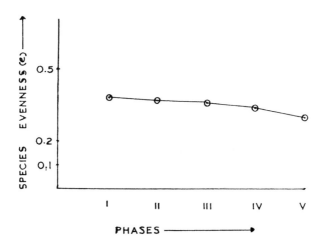

Figure 11.6: Species evenness of wood taxa expressed by phase.

what a randomly selected item would be. Low diversity results when the number of species present is low, or when abundance of each species is variable. Yellen (1977) and Pearsall (1983) have used the Shannon-Weaver information index as a diversity measure and the same index was used in this study. It is seen that the species diversity index attains fairly high values (0.67 in period V to 0.92 in period II). The highest diversity is found in periods II and IV and the lowest in period V.

General diversity is affected by both the number of species present and the evenness with which they are distributed (Potter and Kesselle 1980). To separate these effects Margalef's (1957) measure of species richness and Shannon and Weaver's (1949) index of species evenness were used. Once again period IV (1.82) and period II (1.77) showed highest value of species richness index, the lowest being in period III (1.03). As far as species evenness is concerned all the phases show some constancy in this index with the values ranging from 0.30 in period V to 0.38 in period I.

The above account indicates that the vegetation consisted of stable populations having a large number of species with a very high abundance. However, the species do not appear to have been evenly distributed.

11.5. STANDARD SCORES (Z)

The following results were obtained from data from various species.

Species	Standard scores (Z)				
	I	II	III	IV	V
Rice	0	0.79	1.19	1.39	−0.99
Wheats	−0.54	0.73	−1.26	1.53	−0.46
Barley	−1.31	1.53	0.66	−0.20	−0.68
Oats	−1.89	0.14	0.14	2.16	−0.54
Pulses	−1.63	1.28	−0.46	0.69	−0.12
Prunus spp.	−0.76	1.96	−0.33	−0.65	−0.21
Walnut	−0.62	2.04	0.26	−0.06	−0.06

The values of wheats, oats, pulses, and *Prunus* were obtained using data of all the species collectively. Standard scores of cereals, pulses, and fruit crops are shown in figs. 11.7, 11.8, and 11.9 respectively.

In attempt to convert seed counts to a more usable form, the data was converted to standard scores or the number of standard deviation units by which each taxon varied from the mean. The standard scores provide an idea of the mean count of each species through time and can be examined as data points rather than raw counts. This reduces the impact of absolute quantities and evens out insignificant differences (Blalock 1972). Similarities in the direction of changes obscured by different absolute counts can be very easily seen (Pearsall 1983).

The standard scores of cereal taxa indicate that period IV accounts for high positive deviation from the mean for rice, wheat, and oats whereas period II accounts for high positive deviation for barley. In periods II, III, and V, most of the taxa occur below the mean (fig. 11.7). The standard scores of pulses occur below the mean in periods I, III, and V and above the mean in periods II and IV, the highest deviation being in period II (fig. 11.8).

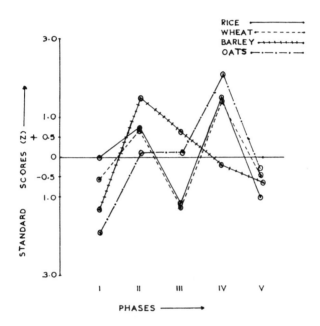

Figure 11.7: Standard scores of cereals expressed as the number of standard deviation units from mean.

The endocarps of *Prunus* show negative deviation from the mean in standard scores in all the periods except period II, where it shows a high positive deviation. Similarly, standard scores for walnut show very high positive deviation in period II and remain around the mean in other periods (fig. 11.9).

Thus, barley, being used more than rice and wheats up to period II, gives way to rice in period IV. At this stage oats also show the highest existence. The data for pulses and fruit crops are not quite significant.

234 *Palaeoethnobotany*

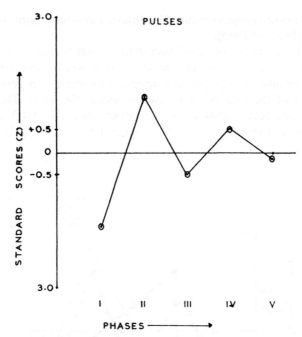

Figure 11.8: Standard scores of pulses.

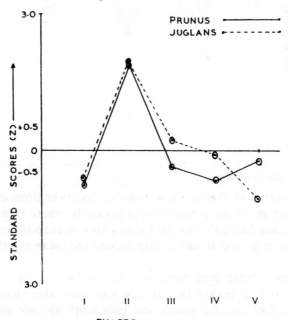

Figure 11.9: Standard scores of fruit crops.

11.6 COEFFICIENT OF SIMILARITY (S)

The following values were obtained for cereal data.

Period	II	III	IV	V
I	0.75	0.67	0.80	0.75
II		0.89	0.80	0.75
III			0.90	0.89
IV				0.80
V				1.00

For pulses, the following values were obtained.

Period	II	III	IV	V
I	0.33	0.50	0.33	0.80
II		0.66	0.75	0.57
III			0.33	0.40
IV				0.57
V				1.00

For fruit crops (*Prunus* spp. and *Juglans*) the following values were obtained.

Period	II	III	IV	V
I	0.80	0.80	0.80	0.50
II		0.50	0.66	0.40
III			1.oo	0.80
IV				0.80
V				1.00

For wood taxa, the following values were obtained.

Period	II	III	IV	V
I	0.50	0.40	0.38	0.24
II		0.32	0.48	0.35
III			0.30	0.25
IV				0.55
V				1.00

11.7 COEFFICIENT OF SIMILARITY (T)

For cereal data, the following values of T were obtained.

Period	II	III	IV	V
I	0.44	0.50	0.56	0.67
II		0.79	0.77	0.56

III		0.58	0.64
IV			0.66
V			1.00

For pulses the following values were obtained.

Period	II	III	IV	V
I	0.22	0.33	0.125	0.29
II		0.36	0.38	0.50
III			0.20	0.22
IV				0.45
V				1.00

For fruit crops, the following values were obtained.

Period	II	III	IV	V
I	0.11	0.31	0.40	0.25
II		0.29	0.11	0.21
III			0.55	0.57
IV				0.44
V				1.00

For wood taxa, the following values were obtained.

Period	II	III	IV	V
I	0.53	0.51	0.41	0.42
II		0.48	0.52	0.54
III			0.37	0.48
IV				0.60
V				1.00

The values of coefficients of similarity provide an insight into the relationship between any pair of periods regarding man-plant interactions and other ecological and subsistence factors. These values reveal how much any two periods share in common. The values of S and T vary from 1, when all the species and their number are common, to zero, when none of the species are common in a pair of sites or phases under consideration.

For cereals, values of S vary from 0.67 (between I and III) to 0.90 (between III and IV). For pulses, S varies from 0.33 (between I & II, I & IV and III & IV) to 0.80 (between I & V). For fruit crops the S varies from 0.40 (between II & V) to 0.80 (between I & II, I & III, I & IV, III & V and IV & V). For wood taxa its values range from 0.26 (between III & V), to 0.55 (between IV & V).

However, this index (S) does not reflect the actual situation as the number of individuals of the species are not taken into consideration. To compare the periods on the basis of quantitative index, values of T become significant.

For cereals the values of T range from 0.44 (between phases I & II) to 0.79 (between II & III). For pulses the values range from 0.125 (between I & IV) to 0.50 (between II & V). For fruit crops the values range from 0.11 (between I & II and II & IV) to 0.57 (between III & V). For wood taxa the values vary from 0.37 (between III & IV) to 0.60 (between IV & V). Corresponding values between any pair of phases give an idea of similarity with respect to a particular group of plants.

11.8. CONCLUDING REMARKS

An attempt has been made to apply palaeoethnobotanical data to the problem of determining stability of subsistence strategies through time and drawing ecological implications on vegetation through the data from Semthan. Some conclusions arrived at could be summed up as follows:

1. Interpretation of charred wood data, analysed first to demonstrate areas of statistical differences and then to show the direction and pattern of these changes, can generate hypotheses to be tested against other archaeological data and further excavations. In the present study it has been hypothesized that during periods I and II a conservative strategy of collection of wood was maintained by substitution of taxa in the face of declining availability of the gymnosperm taxa. At the end of these periods a change in collection strategy occurred with the adoption and collection of more and more broad-leaved elements in exact numbers to the pattern.

2. A site with poor preservation, that is, nothing but charred wood, can still yield interesting data on human interactions with the environment.

3. There is near agreement in all the measures employed for a high degree of constancy from periods I to IV (from 3500 to 1500 B.P.). It is therefore hypothesized that a long-term stable strategy of utilization of resources existed. The relative importance of each taxon is difficult to determine. The slight change in direction in period V is attributed to a slight unevenness of data from this period.

4. Periods IV and II have been the prosperous phases in the cultural evolution.

The statistical analyses of hard archaeobotanical data can yield interesting inferences on human interactions with the environment and the extent to which various taxa of plants were utilized at different stages. However, in this analysis there are some sources of error that might have affected the conclusions drawn. These include smaller counts of charred macroremains of some taxa from some time periods which affect the pattern of distribution in the analysis. Another potential source of error is differential preservation of wood taxa based on their relative hardness.

Table 11.6. Number of grains, seeds, endocarps, or pieces of plant species from different phases at Burzahom and Semthan.

Botanical species	Burzahom				Semthan				
	I	II	III	IV	I	II	III	IV	V
Triticum aestinum	10	18	20	25	10	64	3	71	17
T. sphaerococcum	23	10	9	8	6	–	–	8	–
Oryza sativa	–	–	4	8	29	45	16	57	23
Hordeum vulgare	9	10	11	14	7	144	46	56	27
Avena fatua	–	–	–	–	–	3	2	6	–
A. sativa	–	–	–	–	–	–	1	3	4
Panicum spp.	–	–	–	–	–	–	4	–	–
Setaria spp.	–	–	–	–	–	–	–	3	–
Lens culinaris	1	1	4	5	2	1	–	4	1
Pisum sativum	1	–	1	2	–	4	–	4	1
Phaseolus aureus	–	–	–	–	1	5	2	2	–
P. mungo	–	–	–	–	–	4	2	1	–
P. aconitifolius	–	–	–	–	–	1	–	2	–
Juglans regia	3	4	2	5	2	10	3	2	–
Prunus armeniaca	2	2	–	5	1	9	3	2	7
P. persica	2	–	5	2	–	–	2	1	4
P. amygdalus	–	4	–	1	–	–	–	–	–
P. domestica	–	–	2	–	–	–	–	–	–
P. cerasus	–	–	–	–	–	36	–	–	–
Celtis australis	–	1	2	2	3	4	–	–	1

Statistics in Palaeoethnobotany

Medicago spp.	2	–	1	–	–	–	2	–
Vicia/Lathyrus spp.	1	–	2	4	–	6	9	–
Astragalus spp.	2	–	3	5	–	–	–	–
Galium aparine	3	–	3	7	–	–	11	–
G. tricorne	–	–	–	3	–	–	15	7
G. asperuloides	–	–	–	–	–	–	11	–
Ipomoea spp.	9	4	–	2	–	–	–	–
Melilotus albus	–	1	2	–	–	–	–	1
Lithospermum arvense	–	3	2	4	–	6	10	4
Vitis vinifera	–	2	–	–	–	–	–	–

BIBLIOGRAPHY

Agarwal, V. S. 1970. *Wood yielding plants of India.* Calcutta: Indian Museum.
Agrawal, D. P., S. Kusumgar and R. V. Krishnamurthy (eds.). 1985. *Climate and geology of Kashmir: last four million years.* New Delhi: Today and Tomorrow Printers and Publishers.
Ahmad, S. M. 1986. Central Asia and Kashmir. In G.M. Buth (ed). *Central Asia and Western Himalaya: a forgotten link.* Jodhpur: Scientific Publishers, pp. 1–8.
Alexander, J. 1969. The indirect evidence for domestication. In P. J. Ucko and G. W. Dimbleby (eds.). *Domestication and exploitation of plants and animals.* London: Duckworth, pp. 123-130.
Anderson, E. 1942. Prehistoric maize from Canyon del Muerto. *American Journal of Botany* 29:832-835.
Anonymous, 1969a. *Wealth of India.* New Delhi Publications and Information Director, C.S.I.R.
Anonymous. 1969b. *A handbook of agriculture.* New Delhi: I.C.A.R. Publication.
Arnaudov, N. 1936. Uber prahistorische and subrezente Pflanzenreste aus Balgarien, *TPY-HOBE HA SBJFAPACKOTO IIPNPOIIOM II NTAT-EIIHO IIPYKECTBO KHNT AXVII* Sofia.
*Arnaudov, N. 1937–1938. Untersuchung uber Pflanzenreste aus den Ausgrabungen bei Sadowetz in Balgarien. *Annuaire de l'Universit'e de Sofia, Faculté Physico-Mathematique* XXXIV Livre 3, *Sciences Naturelles* Sofia.
*Arnaudov, N. 1948–49. *Annuaire de l'Universite de Sofia, Faculte de' Sciences* XLV Livre 3, *Sciences Naturelles* Sofia.
Asch, N. B. and D. L. Asch. 1975. Plant remains from the Zimmerman site-grid A. A quantitative perspective. In M. K. Brown. *The Zimmerman site: further excavations at the Grand Village of Kaskakia.* Illinois State Museum. Springfield, pp. 116-120.
Asch, D. L., K. B. Farnsworth and N. B. Asch. 1979. Woodland subsistence and settlement in west central Illinois. In D. S. Brose and N. Greber (eds.). *Hopewell archaeology. Journal of Archaeology* Spec. Pap. 3: 80-85.
Backer, C. A., V. D. Bakhuizen and R. C. Brink. 1968. *Flora of Java* III. Groningen:Noordhoff.
Bakshi, J. S. and R. S. Rana. 1974. Barley. In Sir J. Hutchinson (ed.). *Evolutionary studies in world crops: Diversity and change in the Indian subcontinent.* Cambridge: University Press, pp. 47-54.
Barghoorn, E. S. 1971. The oldest fossils. *Scientific American* 224:30-42.
Barigozzi, C. 1986. *The origin and domestication of cultivated plants.* Amsterdam Elsevier.
Bentham, G. and J.D. Hooker. 1862–1883. *Genera Plantarum.* 3 Vols. London: Reeves & Co.
Ben-ze'ev, N., and D. Zohary 1973. Species relationships in genus *Pisum O. Isreal Journal of Botany* 22:73-91.
*Bertsch, K. and F. Bertsch. 1949. *Geschichte unserer.* Stuttgart: Kulturpflazen.

---

*Original not seen

Bibliography

*Biffen, R. H. 1934. Report on grain from the Fayum. In G. Thompson Caton and E. W. Gardner. *The desert Fayum*. Gloucester:

*Bishop, C. W. 1933. The neolithic age in northern China. *Antiquity* 7:389-404.

Bisht, R. S. 1977. *Banawali: a look back into pre-Indus and Indus civilizations*. Archaeological Survey of India: Pamphlet.

Bisht, R. S. 1986. Reflection on Burzahom and Semthan excavations and later mythological periods in Kashmir valley. In G. M. Buth (ed.). *Central Asia and Western Himalaya: a forgotten link*. Jodhpur: Scientific Publishers, pp. 51-58.

Bisht, R. S. and G. S. Gaur. 1982. *Semthan excavations*. Unpublished ms.

Blalock, H. M., Jr. 1972. *Social statistics*. New York:

Bor, N. L. 1960. *The grasses of Burma, Ceylon, India and Pakistan*. Oxford: Pergamon Press.

Bose, D. M., S. N. Sen and B. V. Subbarayappa. 1971. *A concise history of science in India*. New Delhi: Indian National Science Academy.

Botteima, S. 1984. The composition of some modern charred seed assemblages. In W. Van Zeist and W. C. Casparie (eds.). *Plants and ancient man: Studies in palaeoethnobotany*. Rotterdam: Balkema.

Braidwood, R. J. and B. Howe. 1960. *Prehistoric investigations in Iraqi Kurdistan*. Chicago: Chicago University Press.

Brandis, D. 1971. *Indian trees*. Dehra Dun: Govt. of India Publication.

Brown, H. P. 1925. *An elementary manual on Indian wood technology*, Calcutta: Govt. of India Publication.

Brown, H. P., A. J. Panshin and C. C. Forsaith. 1949. *Textbook of wood technology*, Vol. I. New York: McGraw-Hill Book Co.

Burt, B. C. 1941. Comment on cereals and fruits. In M. S. Vats. *Excavations at Harappa*. New Delhi: Govt. of India Publication, pp. 466.

*Buschan, G. 1895. *Vorgeschichtliche Botanik der Kulturund Nutzpflanzen der alten Welt*. Breseau:

Buth, G. M. 1970. *Investigation of plant remains recovered from archaeological excavation*. Unpublished Ph.D. thesis, Aligarh Muslim University, Aligarh.

Buth, G. M. 1982. SEM study as an aid in identification of caryopses of *Triticum*. *Journal of Economic and Taxonomic Botany* 3:537-540.

Buth, G. M., and R. S. Bisht. 1981. SEM study of ancient wood remains from Kashmir. *Current Science* 50(16): 728.

Buth, G. M., R. S. Bisht and G. S. Gaur. 1982. Investigation of palaeoethnobotanical remains from Semthan, Kashmir. *Man and Environment* 6:41-45.

Buth, G. M., and K. A. Chowdhury. 1971. Plant remains from Atranjikhera Phase III. *Palaeobotanist* 20:280-287.

Buth, G. M., and K. A. Chowdhury. 1973. 4,500 years old plant remains from Egyptian Nubia, *Proceedings of Indian National Science Academy* 38:55-71.

Buth, G. M., and R. N. Kaw. 1985. Plant husbandry in Neolithic Burzahom, Kashmir. *Current Trends in Geology* Vol. VI (Climate and Geology of Kashmir): 109-113.

Buth, G. M., M. Khan and F. A. Lone. 1986a. Antiquity of rice and its introduction in Kashmir. In G. M. Buth (ed.). *Central Asia and Western Himalaya: a forgotten link*. Jodhpur: Scientific Publishers: pp. 63-68.

Buth, G. M., M. Khan and F. A. Lone. 1986b. Retrieval and interpretation of plant remains from archaeological sites. *Man and Environment* 10:45-48.

Buth, G. M., and K. S. Saraswat. 1972. Antiquity of rice cultivation. In A. K. M. Ghouse and M. Yunus (eds.). *Research trends in plant anatomy*. New Delhi: Tata McGraw-Hill, pp. 33-38.

Callen, E. O. 1969. Diet as revealed by coprolites. In D. Brothwell and E. S. Higgs. *Science in Archaeology*. London: Thames and Hudson.

Chalam, G. V., and J. Venkateshwarlu. 1965. *Introduction to agricultural botany in India*. Bombay: Asia Publishers.
Chamberlain, C. J. 1932. *Methods in plant histology*. Chicago: Chicago Press.
Champion, H. G. and S. K. Seth. 1968. *A revised survey of the forest types of India*. Delhi: Govt of India Publication.
Chang, T. T. 1975. Exploration and survey in rice. In O. H. Frankel and J. G. Hawkes (eds.). *Crop genetic resources for today and tomorrow*. Cambridge: pp. 159-165.
Chang, T. T. 1976a. The rice cultures. *Philosophical Transactions of the Royal Society of London* B 275:143-157.
Chang, T. T. 1976b. The origin, evolution, cultivation, dissemination and diversification of Asian and African rices. *Euphytica* 25:425-441.
Chang, T. T. 1976c. *Manual on genetic conservation of rice germplasm for evaluation and utilization*. Los Banos, Philippines: IRRI. 77 pp.
Chang, T. T. 1983. The origins and early cultures of the cereal grains and food legumes. In D. N. Knightley (ed.). *The origins of Chinese civilization*. Berkeley, Los Angeles and London University of California Press, pp. 65-94.
Chang, T. T. 1984. Conservation of rice genetic resources: luxury or necessity? *Science* 224:251-256.
Chang, T. T. 1985a. Crop history and genetic conservation: rice—a case study. *Iowa State Journal of Research* 59(4):425-455.
Chang, T. T. 1985b. The impact of rice on human civilization and population expansion. *Interdisciplinary Science Reviews* 10:4.
*Chang, T. T. 1986a. The ethnobotany of rice in insular southeast Asia. *Asian Perspectives*: (in press).
Chang, T. T. 1986b. Domestication and spread of the cultivated rices. *Recent Advances in the Understanding of Plant Domestication and Early Agriculture*, World Archaeological Congress, Southampton, pp. 1-13.
Chang, T. T., and G. C. Loresto. 1984. The rice remains. In C. Higham and A. Kijngara (eds.). B.A.R. 5231, *Prehistoric investigations in Northeastern Thailand*, Part II. Oxford: B.A.R., pp. 384-385, 390.
Chatterjee, D. 1951. Notes on the origin and distribution of wild and cultivated rices. *Indian Journal of Genetics and Plant Breeding* 11:18.
*Chekiang Provincial Cultural Commission and Chekiang Provincial Museum. 1976. Ho-Mu-tu discovery of important primitive society, an important site. *Wen Wu* 8:6-13. (In Chinese.)
Childe, V.G. 1957. Old World prehistory: Neolithic. In A.L. Koreber(ed) *Anthropology Today: an Encyclopacdia Inventory*. Chicago: Chicago University Press, pp. 193-210.
Chowdhury, K. A. 1934. An improved method of softening hard woody tissues in hydrofluoria acid under pressure. *Annals of Botany* 48:309.
Chowdhury, K. A. 1963. Plant remains from Deh Morasi Ghundai, Afghanistan. *Anthropological Papers American Museum of Natural History*, 50, Part 2, p. 126.
Chowdhury, K. A. 1964. cf. Chowdhury et al. 1977. *Ancient agriculture and forestry in North India*. New Delhi: Asia Publishing House.
Chowdhury, K. A. 1965a. Wood remains from Gen pits of Sabaragamuia Province of Ceylon. *Spolia Zeylanica* 30 Part II:3–6.
Chowdhury, K. A. 1965b. Plant remains from pre- and protohistoric sites and their scientific significance. *Science and Culture* 31:177-178.
Chowdhury, K.A. 1970. Archaeological plant remains from pre- and protohistoric period as a source of history of science. *Indian Journal of History of Science* 5:141-143.
Chowdhury, K. A. 1971. Botany, Prehistoric period. In Bose et al. (eds.). *A concise history of science in India*. New Delhi: Indian National Science Academy, p. 371.
Chowdhury, K. A. 1974a. History of Indian cereals. In K. R. Surange et al. (eds.). *Aspects and appraisal of Indian palaeobotany*, Lucknow: Birbal Sahni Institute of Palaeobotany.

Chowdhury, K. A. 1974b. History of timber plants, fibres and pulses. In K. R. Surange et al. (eds.). *Aspects and appraisal of Indian palaeobotany*, Lucknow: Birbal Sahni Institute of Palaeobotany.
Chowdhury, K. A., and G. M. Buth. 1970. Seed coat structure and anatomy of Indian pulses. *Botanical Journal of Linnean Society* 63:169-177.
Chowdhury, K. A., and S. S. Ghosh. 1946. Report on wood and fruit shells from Arekamedu, Pondicherry. *Ancient India* 2:104-108.
Chowdhury, K. A., and S. S. Ghosh. 1951. Plant remains from Harappa, *Ancient India* 7:3-19.
Chowdhury, K. A., and S. S. Ghosh. 1952. Wood remains from Sisupalgarh. *Ancient India* 8:28-32.
Chowdhury, K. A., and S. S. Ghosh. 1954-55. Plant remains from Hastinapurra (1950-1952). *Ancient India* 10 and 11:121-137.
Chowdhury, K. A., and S. S. Ghosh. 1955. The study of archaeological plant remains and its significance. *Transactions of the Bose Research Institute* 20:79.
Chowdhury, K. A., and S. S. Ghosh. 1957. Charcoal remains, Maski excavation. *Ancient India* 13:133-141.
Chowdhury, K. A., K. S. Saraswat and G. M. Buth. 1977. *Ancient agriculture and forestry in North India*. New Delhi: Asia Publishing House.
Chowdhury, K. A., K. S. Saraswat, S. N. Hassan and R. C. Gaur. 1971. 4000–3500 year old barley, rice and pulses from Atranjikhera. *Science and Culture* 37:531-533.
*Coffman, F. A. 1946. The origins of cultivated oats, *Journal of American Society of Agronomists* 38.
*Coffman, F. A. (ed.). 1961. Oats and oat improvement. *Agronomy*. 8, Wisconsin.
Cohen, M. N. 1977. *The food crisis in prehistory*. New Haven: Yale University Press.
Corner, E. J. H. 1951. The leguminous seed. *Phytomorphology* 1:116-150.
Corner, E. J. H. 1976. *The Seeds of Dicotyledons*. 2 Vols. Cambridge: University Press.
*Costantin, J. and D. Bois. 1910. Sur les grains et tubercules des Tombeaux Peruviens de la Period Incasique. *Revue Generale de Botanique* 22: 242-246.
Costantini, L. 1984. The beginning of agriculture in the Sibbi Kacchi plain and evidence from Mehrgarh. In B. and R. Allchin (eds.). *South Asian archaeology—1981*. Cambridge: Cambridge University Press.
Dao, T. T. 1985. Types of rice cultivation and its related civilization in Vietnam. *East Asian Cultural Studies* 24:41-56.
Darlington, C. D. 1964. *Chromosome botany and the origin of cultivated plants*. London: Allen & Unwin.
Davies, D. R. 1976. Peas. In N. W. Simmonds (ed.). *Evolution of crop plants*, London: Longman. pp. 172-174.
Davis P. H. 1970. *Flora of Turkey*, Vol. 3. Edinburgh: Univ. Press. pp. 370-373.
*De Candolle, A. 1883. *Origine des plants cultivees*. Paris: Librairie Germer Bailliere et cie.
De Candolle, A. 1884. *Origin of cultivated plants* (Reprinted 1959), New York: Haffner.
Dennel, R. W. 1972. The interpretation of plant remains. In E. S. Higgs (ed.). *Papers in economic prehistory*. Cambridge: University Press, pp. 149-159.
de Terra, H. and Paterson, T. T. 1939. *Studies on the Ice Age in India and associated human cultures*. Carnegie Inst. Washington Publ. No. 439, pp. 1-354.
Dikshit, K. N. 1982. The neolithic cultural frontiers of Kashmir. *Man and Environment* VI: 30-36.
Dimbleby, G. W. 1967. *Plants and archaeology*, London: Duckworth.
Dodia, R., D. P. Agrawal and A. B. Vora 1985. New pollen data from the Kashmir dogs: A summary. *Current Trends in Geology*. Vol. VI (*Climate and Geology of Kashmir*): 101-108.
Drew, F. 1875. *The Jammu and Kashmir Territories*. London:
Ellerton, S. 1939. The origin and geographical distribution of *Triticum sphaerococcum* and its cytogenetic behaviour with *T. vulgare*. *Journal of Genetics* 38:307-324.

Erichsen-Brown, C. 1979. *Use of plants for the past 500 years*. Aurora, Canada: Breezy Creeks Press.

Feldman, M. 1976. Wheats. In N.W. Simmonds (ed.). *Evolution of world crops*. London: Longman, pp. 120-128.

Firdaus, A.A. 1944. The afforestation of Shankaracharya hills. *Indian Forester* 70(2):83.

*Fonner, R. L. 1957. *Soc. Amer. Archaeol. Mem.* 14:303-304.

Ford, R. I. 1979. Palaeoethnobotany in American archaeology. In M. Schiffer (ed.). *Advances in archaeological method and theory*. Vol. 2. New York: Academic Press, pp. 286-336.

Gamble, J. S. 1972. *A manual of Indian trees*. London: Sampson Low Marston Company.

Ganju, M. 1945. *Textile industries in Kashmir*. Delhi:

Gaul, J. H. 1948. *The Neolithic period in Bulgaria*. Cambridge, Mass.:

Gaur, G. S. 1987. Semthan excavations: a step towards bridging the gap between the neolithic and the Kushan periods in Kashmir. In B. M. Pande and B. D. Challopadhyaya (eds.). *Archaeology and history: Essays in memory of Sh. A. Ghosh*, pp. 327-337.

Ghose, R. L. M., M. B. Ghatge and V. Subramanyam. 1960. *Rice in India*, New Delhi: I.C.A.R. Publication.

Ghosh, S. S. 1950. Re-examination of a wooden post from Kirari, Madhya Pradesh. *Ancient India* 6:17-20.

Ghosh, S. S. 1961. Further records of rice (*Oryza sativa*) from ancient India. *Indian Forester* 87:295-301.

Ghosh, S. S. and K. Lal 1958. On sal wood excavated from Mauryan Pillared Hall Pataliputra. *Current Science* 27:136-137.

Ghosh, S.S. and K. Lal. 1961. Plant remains from Asvamedha sites near Kalsi of Dehra Dun district. *Science and Culture* 27:188.

Ghosh, S. S. and K. Lal. 1962-63. Plant remains from Rangpur. *Ancient India* 18 and 19:161-175.

Ghouse, A. K. M. 1969. Classification of woods of indigenous coniferous genera of India by microscopic structure. In K. A. Chowdhury (ed.). *Recent advances in Anatomy of tropical seed plants*. New Delhi: Tata McGraw-Hill, pp. 137-150.

Gilmore, M. R. 1931. Vegetal remains of the Ozark Bluff-dweller culture. *Papers of Michigan Academy of Science Arts and Literature* 14:83-102.

Gimbutas, M. 1956. *The prehistory of Eastern Europe*, Part I: *Mesolithic, Neolithic and Copper Age cultures in Russia and Baltic area*. Cambridge: University Press.

Glover, I. C. 1977. The late Stone Age in eastern Indonesia. *World Archaeology* 9:42-61.

Gupta, H. P., C. Sharma, R. Dodia, C. Mandavia and A. B. Vora. 1985. Palynostratigraphy and palaeoenvironment of Kashmir: Hirpur Locality III. *Current Trends in Geology* Vol. VI (*Climate and Geology of Kashmir*):74-90.

Hansen, J., and J. M. Renfrew 1978. Palaeolithic Neolithic seed remains at Franchthi Cave, Greece. *Nature* 271:349-352.

Harlan, J. R. 1968. On the origin of barley. USDA Agriculture Handbook 338. pp. 349-352.

Harlan, J. R. 1971. Agricultural origins: Centres and non-centres. *Science* 174:468-474.

Harlan, J. R. 1976. Barley. In N. W. Simmonds (ed). *Evolution of crop plants*. London: Longman, pp. 93-98.

Harlan, J. R. 1986. Plant domestication: diffuse origins and diffusion. In C. Barigozzi, (ed.). *The origin and domestication of cultivated plants*. Amsterdam: Elsevier, pp. 21-34.

Harlan, J. R., and D. Zohary. 1966. Distribution of wild wheats and barleys. *Science* 153:1074.

*Harms, H. von, 1922. Ubersicht der bisher in altperuanishen Greaten gefundenen Pflanzenreste. In *Eestschrift Eduard Seler*. Stuttgart: Springer Verlag, pp. 157-186.

Harris, D. R. 1977. Alternative pathways towards agriculture. In C. A. Reed (ed.). *Origins of agriculture*. The Hague: Mouton, pp. 179-243.

Harris, D. R. 1984. Ethnohistorical evidence for the exploitation of wild grasses and forbs: Its scope and archaeological implications. In W. Van Zeist and W. A. Casparie (eds.). *Plants and ancient man*. Rotterdam: Balkema.

Harris, D. R. 1986. Plant and animal domestication and the origins of agriculture: The contribution of radiocarbon acceleration dating. In J. A. J. Gowlett, and R. E. M. Hedges (eds.). *The results and prospects of accelerator radiocarbon dating.* Oxford:
Hassan, F. A. 1977. The dynamics of agricultural origins in Palestine: A theoretical model. In C. A. Reed (ed.). *Origins of agriculture.* The Hague: Mouton, pp. 589-609.
Hawkes, J., and L. Woolley. 1963. *History of mankind.* London: Allen and Unwin.
Hector, J. M. 1936. *Introduction to botany of field crops. Vol. I. Cereals.* Johannesburg: Central News Agency.
Hector, J. M. 1937. *Introduction to botany of field crops. Vol. II. Non-cereals.* Johannesburg: Central News Agency.
Heer, O. 1866. Treatise on the plants of the lake dwellings. In F. Keller trans. J. E. Lee. *The lake dwellings of Switzerland and other parts of Europe.* London:
*Helbaek, H. 1938. Planteavl. *Aarboger for Nordisk Oldkyndighed og Historie.* Copenhagen:
Helbaek, H. 1948. Les empreintes de cereales. In P.J. Riis. Copenhagen: Hama
Helbaek, H. 1951. Ukrudtsfr som Naerengsmiddel; forromersk Jernalder. *Kuml* 1951:65-74.
Helbaek, H. 1952a. Spelt (*Triticum spelta* L.) in bronze age Denmark. *Acta Archaeologica* 23:97-107.
Helbaek, H. 1952b. Early crops in southern England. *Proceedings of the Prehistoric Society* XVIII:194.
Helbaek, H. 1952c. Preserved apples and *Panicum* in prehistoric site at Noore Sandegaard in Bornholm. *Acta Archaeologica* 23:108.
Helbaek, H. 1953a. Queen Icetis' wheat. *Dan. Biol. Medd.* 21(8).
Helbaek, H. 1953b. Appendix I. In E. Gjerstad. *Early Rome.* Lund.
Helbaek, H. 1953c. Archaeology and agricultural botany. *Annual Report of Institute of Archaeology* 9:44.
Helbaek, H. 1954a. Store Valby Komavl i. Danmarks Forste Neolitiske Fase. *Aarboger for Nordisk Oldkyndighed og Historic* 202-204.
Helbaek, H. 1954b. Prehistoric food plants and weeds in Denmark. A survey of archaeobotanical research 1923-1954. *Danmarks Geoligishe Unders* II (80):250-261.
Helbaek, H. 1955. The botany of the Vallhagar Iron Age field. In M. Stenberger. *Vallhagar, a Migration Period site on Gotland, Sweden.* Stockholm:
Helbaek, H. 1956. Vegetables in the funeral meals of the pre-urban Rome. Appendix I. In E. Gjerstad. *Early Rome II. Acta Inst. Roman Sucerciae* ser 4. 27(2):287-294.
Helbaek, H. 1958a. Grauballemandens Sidste Maltid. *Kuml* 1958:83-116.
Helbaek, H. 1958b. Plant economy in ancient Lachish. In O. Tufnell (ed). *Lachish.* London: Oxford University Press, pp. 309-317.
Helbaek, H. 1959a. How farming began in the Old World. *Archaeology* 20:3
Helbaek, H. 1959b. Domestication of food plants in the Old World. *Science* 130:365-372.
Helbaek, H. 1959c. Notes on evolution of *Linum. Kuml* 1959.
Helbaek, H. 1960a. The palaeoethnobotany of the Near East and Europe. In R. J. Braidwood, and B. Howe. *Prehistoric investigations in Iraqi Khurdistan.* Chicago: Chicago University Press, pp. 99-118.
Helbaek, H. 1960b. Ecological effects of irrigation in ancient Mesopotamia. *Iraq.* 22:186-196.
Helbaek, H. 1960c. Cereals and weed grasses in phase A. In R. J. Braidwood and L. J. Braidwood. *Excavations in the Plain of Antioch.* Chicago: Bruce.
Helbaek, H. 1960d. Comment on *Chenopodium album* as a food plant in prehistory. *Bericht des Geobotanischen Instituk der Eidg. Techn. Hochschuk, Stiftung Rubel* 31, Heft, 1959. Zurich.
Helbaek, H. 1960e. Ancient crops in the Shahzora valley in Iraqi Khurdistan. *Sumer* XVI:79-81.
Helbaek, H. 1961. Late Bronze Age and Byzantine crops at Beycesultan in Anatolian. *Anatolian Studies* XI:77-97.

Helbaek, H. 1962a. Late cypriote vegetable diet at Apliki. *Opuscula Atheniensia* IV:171-183 Lund.
Helbaek, H. 1962b. Les grains carbonises de la 48-eme couche des fouilles de Tell soukas. *Les Annales Archaeologiques de Syrie* XI-XII.
Helbaek, H. 1963. Palaeoethnobotany. In D. Brothwell, and E. Higgs (eds.). *Science in archaeology*. London: Thames and Hudson, pp. 177-194.
Helbaek, H. 1964a. First impressions of the Catal Huyuk plant husbandry. *Anatolian Studies* XIV:121.
Helbaek, H. 1964b. The Isca grain: A Roman Plant introduction in Britain. *New Phytologist* 63.
Helbaek, H. 1965a. Early Hassunan vegetable food at Tell Es-Sawwan near Samarra. *Sumer* XX:45. Baghdad.
Helbaek, H. 1965b. Isin-Larsan and Horian food remains at Tell Bazmosian in the Dokan valley. *Sumer* XIX:27-35.
Helbaek, H. 1966a. Pre-pottery neolithic farming at Beidha. *Palestine Exploration Quarterly*. 98(1):61.
Helbaek, H. 1966b. Commentary on the phylogenesis of *Triticum* and *Hordeum. Economic Botany* 20:350.
Helbaek, H. 1966c. Vendeltime farming products at Eketorp on Oland, Sweden. *Acta Archaeologica* 37:216-221.
Helbaek, H. 1966d. The plant remains from Nimrud. In M. E. Mallowan. *Nimrud and its remains*. London: Collins, pp. 615.
Helbaek, H. 1966e. What farming produced at Cypriote Kalopsidha. In P. Astrom. *Excavations at Kolopsidha and Ayios Iakovos in Cyprus. Studies in Mediterranean Archaeology*, Vol. II. Lund.
Helbaek, H. 1969. Plant collecting, dry farming and irrigation in prehistoric Deh Luran. In F. Hole et al. *Pre-history and human ecology of the Deh Luran Plain: An early village sequence from Khuzistan Iran*. Memoirs of Museum of Anthropology, University of Michigan. No. 1:244-383.
Helbaek, H. 1970. The plant husbandry of Hacilar. In J. Mellaart. *Excavations in Hacilar, I*. Edinburgh: University Press, pp. 189-244.
Helbaek, H. 1972. Plant anatomy in culture-historical research. In A. K. M. Ghouse and M. Yunus. *Research trends in plant anatomy*. Delhi: Tata McGraw-Hill, pp. 19-32.
Higham, C., and A. Kijngam. 1984. The excavation of Ban Na di Bang Muang Phruk and Non Kao Noi. In C. Higham and A. Kijngam (eds.). B.A.R. S231, *Prehistoric investigations in Northeastern Thailand* Part I. Oxford: BAR, pp. 22-56.
Hillman, G. C. 1972. The plant remains. In D. H. French, G. C. Hillman and S. Payne. Excavations at Can Hasan III in Anatolia, 1969-1970. In E. S. Higgs (ed.). *Papers in economic prehistory*. Cambridge: University Press.
Hillman, G. C. 1975. The plant remains from Tell Abu Hureyra: a preliminary report. In A. M. T. Moore. The excavations of Tell Abu Hureyra in Syria: a preliminary report. *Proceedings of the Prehistoric Society* 41:70-73.
Hillman, G. C. 1978. On the origins of domestic rye—*Secale cereale. Anatolian Studies*. 28:157-174.
Hillman, G. C. 1981. Reconstructing crop husbandry practices from charred remains of plants. In R. Mercer (ed.). *Farming practice in British Prehistory*. Edinburgh: University Press, pp. 123-162.
Hillman, G. C. 1984a. Interpretation of archaeological plant remains: the application of ethnographic models from Thrkey. In W. Van Zeist and W. A. Casparie (eds.). *Plants and ancient man*. Rotterdam: Balkema, pp. 1-42.
Hillman, G. C. 1984b. Traditional husbandry and processing of archaic cereals in recent times: the operations, products and equipment which might feature in Sumerian texts. Part I: the glume wheats. *Bulletin on Sumerian Agriculture* 1:114-152.

Hillman, G. C. 1986. Plant foods in ancient diet: the archaeological role of palaeofacies in general and lindow Man's gut contents in particular. In I. M. Stead, J. Bourke and D. R. Brothwell (eds.). *Lindow Man: The body in the log.* London: British Museum.
Hillman, G. C., S. M. Colledge, and D. R. Harris. 1986a. Plant food economy during the epipalaeolithic period at Tell Abu. Hureyra Syria: dietary diversity, seasonality and modes of exploitation. *Recent Advances in the Understanding of Plant Domestication and Early Agriculture.* World Archaeological Congress, Southampton.
Hillman, G. C., and S. M. Davies. 1986. Experimental measurement of domestication rates in populations of wild einkorn (*Triticum boeoticum*) under primitive system of cultivation. Forthcoming.
Hillman, G. C., E. Madeyska and J. Hather. 1986b (in press) Plant food economy at late Palaeolithic Wadi Kubbaniya in Upper Egypt. In F. Wendorf, R. C. Schild A. Close et al. (eds.). Title as yet unknown.
Hilu, K. W., J. De Wet and J. R. Harlan. 1979. Archaeobotanical studies of *Eleusine coracana* ssp. *coracana* (finger millet). *American Journal of Botany* 66(3):330-333.
Holden, J. H. W. 1966. Species relationships in the Avenae. *Chromosome (Berl.)* 20:75-124.
Holden, J. H. W. 1976. Oats. In N. W. Simmonds (ed.). *Evolution of world corps.* London: Longman, pp. 86-90.
Hole, F., and H. Flannery 1967. The prehistory of southwestern Iran: a preliminary report. *Proceedings of the Prehistoric Society* XXXIII:147, London.
Hopf, M. 1955. Form veranderunga van Getreide-Kornern beim Verkohlen. *Berischte der Deutschen Bontanischen Geselleshasaft* 68.
*Hopf, M. 1957. Botanik und Vorgeschichte. *Jahrbuch Rom. German Zentralmuseums Mains* 4.
*Hopf, M. 1962. Bericht uber die Utersuchung von Samen und Holzkoheresten von der Argissa-Maghula aus der Prakeramischen bis mitteebronzezeitlochen. Schichten. In V. Milojcic, J. Boessneck and M. Hopf. *Die Deutschen Ausgrabungen* auf der *Argissa-Maghula in Thessalien.* Bonn:
Hopf, M. 1969. Plant remains and early farming in Jericho. In G. W. Dimbleby, and P. Ucko. *The domestication and exploitation of plants and animals.* London: Duckworth.
*Hopf, M. 1970. Neolithische Getreide du funde in der hohle von Nerha (Prov Malaya). *Madrider Mitteilumgen.* II:18-34.
Hopf, M. 1986. Archaeological evidence of the spread and use of some members of the leguminosae family. In C. Barigozzi (ed.). *The origin and domestication of cultivated plants.* Amsterdam: Elsevier, pp. 35-60.
Hudson, P. S. (trans.). 1962. *Cultivated plants and their wild relatives.* Farnham, Bucks: Commonwealth Agricultural Bureau.
Hutchinson, J. 1965. *Essays on crop plant evolution.* Cambridge: University Press.
Hutchinson, J., J. G. G. Clark, R. M. Jope and R. Riley. 1977. The early history of agriculture. *Philosophical Transactions of the Royal Society of London* B 275:1-23.
Jane, F. W. 1956. *The structure of wood.* New York: The Macmillan Company.
Jarriage, J. F., and R. H. Meadow. 1980. The antecedents of civilization in the Indus valley. *Scientific American* 243(2):122-132.
Jessen, K., and H. Helbaek. 1944. *Cereals in Great Britain and Ireland in prehistoric and early historic times.* Kgl. Dan. Vidensk Selsk. Biol. Skrifter. Copenhagen:
Johannessen, S. 1981a. Plant remains from Sandy Ridge farm site. In *Final report on archaeological investigations at the Sandy Ridge farm site (11-S-660).* FA 1-270. Archaeological Mitigation Project Report No. 20. Univ. Illinois, Urbana. pp. 18-23.
Johannessen, S. 1981b. Floral resources and remains. In *The Droyff Levin: A late Arehaic occupation in the American Bottom.* Thomas E. Emerson. FAI-270. Archaeological Mitigation Project Report No. 24. University of Illinois, Urbana.

*Jones, V. H. 1936. The vegetal remains of Newt Kash Hollow Shelter. In W. S. Webb and W. D. Frunkhouser, *Rock shelters in Menifee County, Kentucky*. Lexington:

Kachroo, P. 1986. *Archaeological, palaeoclimatic and floristic relations between Kashmir and Central Asia*. Unpublished ms.

Kachroo, P., and M. Arif. 1970. *Pulse crops of India*, New Delhi: I.C.A.R. Publication.

Kajale, M. D. 1974a. Ancient grains from India. *Bulletin of the Deccan College and Research Institute* 34:35-75.

Kajale, M. D. 1974b. Plant economy at Bhokardan. In S. B. Deo, and R. S. Gupta *Excavations at Bhokardan* Nagpur: Marathwada University, pp. 217-223.

Kajale, M. D. 1977a. Plant economy at Inamgaon. *Man and Environment* 1:54-56.

Kajale, M. D. 1977b. On the botanical findings from excavations at Daimabad—a chalcolithic site in Western Maharashtra, India. *Current Science* 46(23):819-820.

Kajale, M. D. 1977c. Ancient grains from excavations at Nevasa, Maharashtra, India. *Geophytology* 7(1):98-106.

Kajale, M. D. 1979. Plant Remains. In S. B. Deo et al. (eds.). *Apegaon Excavations*. Poona: Deccan College.

Kajale, M. D. 1986. *The exploitation of wild plants in Mesolithic period during c. 10500–8000 B.C. in Sri Lanka: Palaeobotanical study on cave site at Beli-Lena (Kitugala)*. World Archaeological Congress, Southampton.

*Kihara, H. 1924. Cytologische und genetische Studien bei Wichtizen Getreidearten mit besonderer Rucksicht auf das Verhalten der Chromosomen und die Sterilitat in den Bastarden. *Memories of College of Science Kiojoto Sea B I*:1-200.

*Kihara, H. 1944. Die Entdeckung des DD—Kyoto. Analysators beim Weizen Volranfige. Mittleilung. *Agriculture and Horticulture* 19:889-890.

Kihara, H. 1954. Considerations on the evolution and distribution of *Aegilops* species based on the analyser method. *Cytologia* 19:336-357.

Kihara, N. 1959. The origin of cultivated rice. *Kihara Institute of Biological Research* 10:68.

Korber-Grohne, U. 1981. Distinguishing prehistoric grains of *Triticum* and *Secale* on the basis of their surface patterns using scanning electron microscopy. *Journal of Archaeological Science* 8:197-204.

Korber-Grohne, U., and U. Peining. 1980. Microstructure of the surfaces of carbonized and uncarbonized grains of cereals as observed in scanning electron and light microscopes as an additional aid in determining prehistoric findings. *Flora (Jena)* 170:189-228.

Kosambi, D. D. 1965. *The culture and civilization of ancient India in historical outline*. London:

*Kunth, C. 1826. Examen botanique, In J. Passalcqua. *Catalogue raisonne et historique des antiquities decouvertes en Egypte*. Paris:

Ladizinisky, G. 1971. Chromosome relationship between tetraploid (2n = 28) *Avena murphyi* and some diploid, tetraploid and hexaploid species of oats. *Canadian Journal of Genetics and Cytology* 13:203-209.

Ladizinisky, G. and D. Zohary. 1971. Notes on species determination, species relationships and polyploidy in *Avena*. *Euphytica* 20:380-395.

Lamberg-Karlovsky, C. C. and T. W. Beale. 1986. *Excavations at Yayha, Iran 1967–1975: the early periods*. American School of Prehistoric Research Bull. 38. Peabody Museum, Harvard.

Lambert, W. J. 1933. *List of trees and shrubs for the Kashmir and Jammu forest circles*. Srinagar: Jammu and Kashmir State.

Langenheim, J. H. and K. V. Thimann. 1982. *Botany: Plant biology and its relation to human affairs*. New York: John Wiley & Sons, Inc.

Lawrence, W. H. 1967. *The valley of Kashmir*. Srinagar:

*Lee, J. E. 1866. *The lake dwellings of Switzerland and other parts of Europe*. London:

Legge, A. J. 1977. The origins of agriculture in the Near East. In J. V. S. Megaw (ed.). *Hunters, gatherers and first farmers beyond Europe*. Leicester: University Press.

Li, H. L. 1970. The origin of cultivated plants in southeast Asia. *Economic Botany* 24(1):3-19.
*Li, K. C. 1983. Problems raised by K'en-ting excavations of 1977. *Bulletin of Department of Archaeology and Anthropology, National Taiwan University* 43:86-116.
*Lissitsina, G. N. 1970. Plantes satives du Proche Orient et du sud de l'Asie Centrale awa VIIIe-Ve millenaires avant notre ere *COBET (KA) APXEOJIO* 3 Moscow.
Lone, F. A., M. Khan and G. M. Buth. 1986a. Beginnings of agriculture in India: An apraisal of the palaeobotanic evidence. *Recent Advances in the Understanding of Plant Domestication and Early Agriculture*. World Archaeological Congress, Southampton.
Lone, F. A., G. M. Buth and M. Khan. 1986b. Wood remains from archaeological sites. *Man and Environment* X:73-78.
Lone, F. A. 1987. *Palaeo-ethnobotanical studies of archaeological sites of Kashmir*. Unpublished Ph.D. Thesis, Kashmir University.
Lone, F. A., M. Khan, and G. M. Buth. 1987. Plant remains from Banawali, Haryana. *Current Science* 56(16):837-838.
Lone, F. A., M. Khan, and G. M. Buth. 1988. Five thousand years of vegetational changes in Kashmir: the impact of biotic factor. *Proceedings of the Indian National Science Academy* 54 A(3):497-500.
Luthra, J. C. 1936. Ancient wheat and its variability. *Current Science* 4(7):459.
Mackay, J. 1954. The taxonomy of hexaploid wheat. *Svensk Botanisk Tidskrift* 48.
Mangelsdorf, P. C., and C. E. Smith, Jr. 1949. New archaeological evidence on the evolution of maize. *Botanical Museum Leaflet* 13:213-247. Harvard University.
Mangelsdorf, P. C., R. S. Macneish and W. C. Galinat. 1956. Archaeological evidence on the diffusion and evolution of maize in Mexico, *Botanical Museum Leaflet* 17:125-150. Harvard.
Margalef, R. 1957. Information theory in ecology. *General Systems Bulletin* 31:36-71.
Martin, A. C., and W. D. Barkely. 1961. *Seed identification manual*. University of California Press.
Martin, J. H., and W. H. Leonard. 1967. *Principles of field crop production*. New York:
*Matthias, W. and J. Schultze-Motel. 1967. Kulturpflanzenabdrucke an Schnurkeramischen Gefasen aus mittle deust chland. *Jahresschrift für mitteldeulsche Vergeschichte* 51 Halle
McFadden, E. S., and E. R. Sears. 1946. The origin of *Triticum spelta* and its free threshing hexaploid relatives. *Journal of Heredity* 37(3):81-90; (4):107-116.
Mehta, M.L. 1947. The afforestation of Shankaracharya hills. *Indian Forester* 73(6):270.
Meldgaard, J., P. Mortensen and H. Thrane. 1963. Excavations at Tepe Guran, Luristan. *Acta Archaeologica* XXXIV.
Miles, A. 1978. *Photomicrographs of world woods*. New York: Department of Environment.
Miller, N. 1984. The interpretation of some carbonised cereal remains. *Bulletin of Sumerian Agriculture* 1: 45-47.
Minnis, P. E. 1981. Seeds in archaeological sites: sources and some interpretive problems. *American Antiquity* 46(1): 143-152.
Mirza Haider, D. 1973. *Tarikh-i-Rashidi*. Patna: E & R.
Moore, A. M. 1975. The excavations of Tell Abu Hureyra in Syria: a preliminary report. *Proceedings of Prehistoric Society* 41:50-77.
Moore, A. M. T. 1982. Agricultural origins in the Near East: A model for the 1980's. *World Archaeology* 14(2): 224-236.
Moore, A. M. T. 1985. The development of Neolithic societies in the Near East. *Advances in World Archaeology* 4:1-69.
Morishima, H., K. Hinata and H. I. Oka, 1963. Comparison of modes of evolution of cultivated forms from two wild rice species *O. breviligulata* and *O. perennis*. *Evolution* 17:170-181.
Musil, A. 1963. *Identification of crop and weed seeds*. Agriculture Handbook No. 219, Washington, D.C: U.S. Department of Agriculture.

*Natho, I. and W. Rothmaller. 1957. Bandkeramische Kulturpflanzenreste aus Thuringen und Sachsen. Bertrage Z.Fd.L. III:73-98. Berlin.

Nayar, N. M. 1973. Origin and cytogenetics of rice. *Advanced Genetics* 17:153-292.

Neuweiler, E. 1905. Die Prahistorischen Pflanzenreste Mitteleuropas. *Vierteljahresschrift der naturforschenden Gesellschaft* in Zurice.

Nilan, R. A. (ed.) 1971. Barley Genetics II. *Proceedings of Second International Barley Genetics Symposium*. Washington.

Noy, T., A. J. Legge and E. S. Higgs. 1975. Recent excavations at Nahal Oren, Israel. *Proceedings of the Prehistoric Society* 39:75-99.

Oka, H. I. 1974. Experimental studies on origin of cultivated rice. *Genetics* 78:475-486.

Olmo, H. P. 1976. Grapes, *Vitis, Muscadinia* (Vitaceae). In N. W. Simmonds, *Evolution of crop plants*. London: Longman, pp. 294-298.

Panshin, A. J., C. De Zeeu and H. P. Brown. 1964. *Textbook of wood technology*, Vol. I. New York: McGraw-Hill.

Pant, R. K. C. Gaillard, V. Nautiyal, G. S. Gaur and S. L. Shali. 1982. Some new lithic and ceramic industries from Kashmir. *Man and Environment* 6:37-40.

Parker, R. N. 1924. *A forest flora for the Punjab with Hazara and Delhi*, Lahore:

Pearsall, D. M. 1979. The application of ethnobotanical techniques to the problem of subsistence in the Ecuadorian Formative. Ph.D. dissertation (Anthropology) University of Illinois.

Pearsall, D. M. 1980. Pachamachay ethnobotanical report: Plant utilization at a hunting base camp. In A. Higgs and J. W. Rick, *Prehistoric hunters of the High Andes*. New York: Academic Press, pp. 191-231.

Pearsall, D. M. 1983. Evaluating the stability of subsistence strategies by use of palaeoethnobotanical data. *Journal of Ethnobiology* 3(2):121-137.

Pearsall, D. M. 1986. *Adaptation of early hunter gathers to the Andean environment*. World Archaeological Congress Papers. Southampton.

Pearson, R. S. and H. P. Brown. 1932. *Commercial timbers of India*, Vols. I and II. Calcutta:

Percival, J. 1921. *The wheat plant*. London: Duckworth.

Peterson, R. F. 1965. *Wheat: Botany, cultivation and utilization*. New York: Interscience Publishers Inc.

Phillips, E. W. J. 1948. Identification of soft woods by their microscopic structure. *Forest Products Research Bulletin* 22:1-56.

Pielou, E. C. 1969. *An introduction to mathematical ecology*. New York:

Porteres, R. 1950. Vieilles agricultures de l'Afrique intertropicale. Centres d'origine et de diversification varietale primaire et bercaux d'agriculture anterieurs au XVI Siecle. *L'agronomie tropicale* V: 489-507.

Potter, M. W. and S. P. Kesselle. 1980. Predicting mosaics and wildlife diversity resulting from fire disturbance to the forest ecosystem. *Environmental Management* 4:247-254.

Prakash, U. and N. Awasthi. 1957-59. In H. D. Sankalia, S. B. Deo and Z. D. Ansari (eds.). *Chalcolithic Excavations at Navdatoli 1957-59*, pp. 440-448.

Puri, G. S. 1948. The flora of the Karewa series of Kashmir and its phytogeographical affinities with chapters on the methods of identification. *Indian Forester* 74(3):105-122.

Puri, G. S. 1957. Preliminary observations on the phytogeographical changes in the Kashmir valley during the Pleistocene. *Palaeobotanist* 6:6-18.

Puri, G. S., V. M. Meher-Homji, R. K. Gupta, and S. Puri. 1983. *Forest Ecology*, Vol. I. New Delhi: Oxford and IBH Pub. Co.

Purseglove, J. W. 1974. *Tropical crops. Vol. I: Monocotyledons*. London: Longman

Purseglove, J. W. 1977. *Tropical crops Vol. II: Dicotyledons*. London: Longman.

Radziniski, W. 1979. *A history of China*, Vol. I. Oxford:

Raikes, R. L. and R. H. Dyson. 1961. The prehistoric climate of Baluchistan and the Indus Valley. *American Anthropology* 63:265-281.

Raizada, M. B. and K. Sahni. 1960. Living gymnosperms. *Indian Forest Records* (N. S.) 512.
*Rajathy, T. and R. S. Sadasivaiah. 1969. The cytogenetic status of *A. magna*. *Canadian Journal of Genetics and Cytology* 11:77-85.
Rajathy, T., and H. Thomas. 1974. Cytogenetics of oats (*Avena*). *Genetic Society of Canada* Misc. Publ. 2:90.
Rao, A. T. 1960. A further contribution to the flora of Jammu and Kashmir State. *Bulletin of Botanical Survey of India* 2(3 and 4):387-423.
Rao, K. R. and R. Sahi. 1967. Plant remains—Prakash. *Ancient India* 20 and 21:139-153.
Rao, M. V. 1974. Wheat. In J. Hutchinson (ed.). *Evolutionary studies in world crops: Diversity and change in the Indian subcontinent*. Cambridge: University Press.
Ray, S. C. 1957. *Early history and culture of Kashmir*. Calcutta:
Reed, C. A. 1976. Discussion and some conclusions. *The origins of agriculture*. The Rague: Mouton.
Reed, C. A. 1977. A model for the origin of agriculture in the Near East. In C. A. Reed (ed.). *The origins of agriculture*. The Hague: Mouton.
Renfrew, J. M. 1965. Appendix VI: Grain impressions from the Iron Age sites of Wandleburg and Balley. In M. D. Craster. *Aldwick, Barley: Recent work at the Iron Age site. Proceedings of the Cambridge Antiquarian Society* LVIII.
Renfrew, J. M. 1966. A report on recent finds of carbonized cereal grains and seeds from prehistoric Thessaly. *Thessalika* 5:21.
Renfrew, J. M. 1968. A note on the neolithic grain from Can Hassan. *Anatolian Studies* XVIII.
Renfrew, J. M. 1969. The archaeological evidence for the domestication of plants: methods and problems. In P. Ucko and G. W. Dimbleby. *The domestication and exploitation of plants and animals*. London: Duckworth.
Renfrew, J. M. 1973. *Palaeoethnobotany: The prehistoric food plants of the Near East and Europe*. London: Methuen & Co.
Richard, L. C. 1819. *Observations on the structure of fruits and seeds*. London: John Harding Publ.
Richaria, R. H. 1960. Origins of cultivated rice. *Indian Journal of Genetics and Plant Breeding*. 20:1-14.
Richaria, R. H. and S. Govindaswami, 1966. *Rices of India*. Patna: Govt. of India Publication.
Riley, R. 1965. Cytogenetics and the evolution of wheat. In J. Hutchinson. *Essays on crop plant evolution*. Cambridge: University Press.
Rindos, D. 1984. *The origins of agriculture: an evolutionary perspective*. New York: Academic Press.
*Royle, J. F. 1839. *Illustrations of the botany and other branches of history of the Himalayan Mountains and of the flora of Cashmere*.
Saffary, D. R. 1876. Les antiquites peruviennes a l' exposition de Philadelphia. *La Nature* 4:401-407.
Safford, W. E. 1917. Food plants and textiles of ancient America. *Proceedings of 9th International Congress of Americanist*, pp. 12-30.
*Sahara, M. and T. Satto. 1984. The earliest rice cultures of India. *Archaeology Journal* (Japan) 228:31-34.
Sahni, B. 1936. The Himalayan uplift since the advent of man: its culthistorical significance. *Current Science* 5(1):57-61.
Sakamura, T. 1918. Kurz Mitteilung uber die Chromosomenzahlen und der Verwandt Sehaftsverhal-tnisse der *Triticum*-arten. *Botanical Magazine Tokyo* 32:151-154.
*Sankalia, H. D., B. Subbarao and S. B. Deo. 1953. *Southwest Journal of Anthropology* 9:343.
Sankalia, H. D. 1971. *Some aspects of prehistoric technology in India*. New Delhi: Deccan College of Research.
Sankalia, H. D. 1974. *Pre- and proto-history of India and Pakistan*. Poona: Deccan College of Research.

Schiemann, E. 1951. New results on the history of cultivated cereals. *Heredity* 5 Part 3:305-318.
Schultze-Motel. J. 1968. Literature uber archaeologische Kulturpflanzenreste (1965-67). *Die Kulturpflanze* XVI.
Schultze-Motel, J. 1980. Literature uber archaeologische Kulturpflanzenreste (1978/79). *Die Kulturpflanze* XXVIII:361-378.
Schultze-Motel, J. 1981. Literature uber archaeologische Kulturpflanzanreste (1979/80). *Die Kulturpflanze* XXIX S. 447-463.
Schultze-Motel, J. 1982. Literature uber archaeologiche Kulturpflanzenreste (1980/81). *Die Kulturpflanze* XXX S. 255-272.
*Sears, E. R. 1956. The systematics, cytology and genetics of wheat. *Handbuch der Pflanzenzuchtung*. Berlin.
Second, G. 1982. Origin of the genic diversity of cultivated rice (*Oryza* spp.): study of the polymorphism scored at 40 isozyme loci. *Japanese Journal of Genetics* 57:25-57.
Sen Gupta, P. R. 1985. Scientific reconstruction of archaeological evidence of Ganga Valley. *Current Science*. 54(2):74-79.
*Sen 1943. cf. C. R. Metcalf and L. Chalk. *Anatomy of the dicotyledons*. Oxford: Claredon Press, 1950.
Shali, S. L. 1986. Neolithic culture in Kashmir with reference to Central Asia. In G. M. Buth, (ed.). *Central Asia and Western Himalaya: a forgotten link*. Jodhpur: Scientific Publishers, pp. 19-26.
Shannon, C. E. and N. Weaver. 1949. *The mathematical theory of communication*. Urbana: University of Illinois Press.
Sharma, A. K. 1982. Excavation at Gofkral 1979-80. *Puratattva*, 11.
Sharma, B. D., and Vishnu-Mittre. 1968. Studies of post-glacial vegetational history from the Kashmir Valley—2. Baba Rishi and Yus Maidan. *Palaeobotanist* 17(3):231-243.
Sharma, C., and H. P. Gupta. 1985. Palynostratigraphy and palaeoenvironments: Krachipatra Lower Karewa Kashmir. *Current Trends in Geology* Vol. VI (*Climate and Geology of Kashmir*):91-96.
Sharma, C., H. P. Gupta, R. Dodia and C. Mandavia. 1985. Palynostratigraphy and palaeoenvironments: Dubjan Lower Karewa Kashmir. *Current Trends in Geology Vol. VI (Climate and Geology of Kashmir*):69-74.
Sharma, G. R., and D. Mandal. 1980. Excavations at Mahagara 1977-78 (A Neolithic settlement in the Belan valley). *Archaeology of the Vindhyas and the Ganga Valley* 6:Allahabad: Allahabad University.
Sharma, G. R. and V. D. Misra, 1980. *Excavations at Chopani Mando (Belan Valley) 1977-79. Epipalaeolithic to protoneolithic*. Allahabad: Allahabad University
*Shaw, F. J. P. 1943. Vegetation remains. In E. J. H. Mackay, *Chanhudaro excavations 1935-36*. Connecticut: New Heaven, pp. 250-251.
Singh, G. 1963. A preliminary survey of the post-glacial vegetational history of the Kashmir valley *Palaeobotanist* 12(1):73-108.
Singh, G. 1971. The Indus Valley culture. *Archaeology and Physical Anthropology in Oceanica* 6(2):
Singh, G. and D. P. Agrawal. 1976. Radiocarbon evidence for deglaciation in north-western Himalaya, India. *Nature* 26:232.
Singh, G. and P. Kachroo. 1976. *Flora of Srinagar and plants of neighbourhood*, Dehra Dun: A Bisher Singh Mahendra Pal Singh.
*Singh, R. D. 1946. *Indian Journal of Genetics and Plant Breeding* 6:34.
Smith, C. R., Jr. 1965. The archaeological records of cultivated crops of New World origin. *Economic Botany* 19:322-334.
Smith, P. M. 1976. Minor crops. In N. W. Simmonds (ed.). *Evolution of crop plants*, London: Longman, pp. 301-324.

Stapf, O. 1931. Comment on cereals and fruits. In J. Marshall (ed.). *Mohenjodaro and the Indus civilization*. London: Arthur Probsthain, p. 586.

Stemler, A., and R. H. Falk. 1980. A scanning electron microscopic study of cereal grains from Wadi Kubbanya. Appendix 3. In A. E. Close (ed.). *Loaves and fishes: The pre-history of Wadi Kubbaniya*. 1.

Stemler, A. and R. H. Falk. 1984. Evidence of grains from the site of Wadi Kubbaniya (Upper Egypt). In L. Krzyzaniak and M. Kobusiewicz. *Origin and development of food producing cultures in North Eastern Africa*. Poznan: Polish Academy of Sciences.

Stewart, R. R. 1972. *Flora of West Pakistan: An annotated catalogue of vascular plants of West Pakistan and Kashmir*. Karachi: Fakhiri Press.

Struever, S. 1968. Flotation technique for the recovery of small-scale archaeological remains. *American Antiquity* 33:353-362.

Struever, S. (ed.). 1971. *Prehistoric agriculture*. New York: Natural History Press.

Tackholm, V., G. Tackholm and M. Drar. 1941. *Flora of Egypt*. Cairo:

Takahashi, R. 1955. The origin and evolution of cultivated barley. *Advanced Genetics* VII:227.

Team of Luo-Jia-Jiao Site. 1981. Excavations at Luo-jia-Jiao site in Tong Xiang County of Zhejiang. *Bull. Zhejiang Provincial Archaeological Institute*, pp. 1-44.

Tempir, Z. 1964. Beitrage Zur alterten geschichte des pflanzenbaus in Ungarn. *Acta Archaeologica Hungaricae* XVI:1-2.

Thapar, R. 1966. *A history of India*. London:

Thibodeau, F. R., and N. H. Nickerson. 1985. Change in wetland plant associations induced by impoundment and drainage. *Biological Conservation* 30:269-278.

Thiebault, S. 1986. Palaeoenvironment and ancient vegetation of Baluchistan based on charcoal analysis of archaeological sites. *International Symp. Palaeoclimatic Palaeoenvironmental Changes in Asia: last four million years*: Abstract 22 Ahmadabad.

Towle, M. A. 1961. *The ethnobotany of pre-columbian Perue*. New York: Viking Fund Publications in Anthropology.

Troup, R. S. 1921. *Silviculture of Indian trees*, Vols. I, II and III. Oxford: University Press.

Tutin, T.G., V.H. Heywood, N.A. Buges, D.H. Valentine, S.M. Watters and D.A. Webb (eds). 1964. *Flora Europea*. Cambridge: University Press, pp. 304.

Ucko, P. J. and G. W. Dimbleby (eds.). 1969. *The domestication and exploitation of plants and animals*. London: Duckworth.

Van Zeist, W. and S. Botteima. 1966. Palaeobotanical investigations at Ramad, *Annales Archaeologiques Arabes Syriennex* XVI:179-180.

Van Zeist, W. and W. A. Casparie. 1968. Wild einkorn wheat and barley from Tell Mureybit in Northern Syria, *Acta Botanica Neerlandica* 17(1): 44-53.

Van Zeist W. and W.A. Casparie. (eds). 1984. *Plants and Ancient Man : Studies in Palaeoethobotany*. Rotterdam:Balkema.

Vats, M. S. 1941. *Excavations at Harappa*, Vol. I. New Delhi: Govt. of India Publication.

Vavilov, N. I. 1949-50. The origin, variation, immunity and breeding of cultivated plants. *Chronica Botanica* 13:1-366.

Vishnu-Mittre. 1961. Plant economy in ancient Maheshwar. In *Technical Report on Archaeological Remains*. No. 2, Deccan College, Poona, pp. 13-52.

Vishnu-Mittre. 1962. Plant economy at ancient Navdatoli. In *Technical Report on Archaeological Remains*. Deccan College, Poona: pp. 11-52.

Vishnu-Mittre. 1963. Oaks in the Kashmir valley with remarks on their history. *Grana Palynologica* 4(2):306-312.

Vishnu-Mittre. (1965). Floristic and ecological reconsiderations of the Pleistocene plant impressions from Kashmir. *Palaeobotanist*, 13(3): 308-327.

Vishnu-Mittre. 1966. Plant remains from Burzahom, Kashmir, in T. N. Khazanchi (ed.). *Excavations of Burzahom*. New Delhi: Govt. of India.

Vishnu-Mittre. 1968a. Protohistoric records of agriculture in India. *J. C. Bose Endowment Lecture. Transactions of the Bose Research Institute, Calcutta* 31.
Vishnu-Mittre. 1968b. *History of agriculture in India*. Poona: Deccan College and Post-Graduate Research Institute.
*Vishnu-Mittre. 1968c. Plant economy in ancient Ahar. In H. D. Sankalia et al. *Copper Age city of Tambavati*, Udhampur. Poona: Deccan College.
*Vishnu-Mittre. 1968d. In M. G. Dikshit (ed.). *Excavations at Kaundinyapur*. Bombay:
Vishnu-Mittre. 1969. Remains of rice and millet. In H. D. Sankalia, S. B. Deo and Z. D. Anari, excavations at Ahar, 1961-62 Poona: Deccan College, pp. 229-236.
Vishnu-Mittre. 1971. In M. S. Nagaraja Rao (eds.). *Protohistoric cultures of the Tungabhadra Valley, Dharwar, India*. Deccan College Poona.
Vishnu-Mittre. 1972a. Neolithic plant economy at Chirand, Bihar. *Palaeobotanist* 22(1):18-22.
Vishnu-Mittre. 1972b. Palaeobotany and the environment of early man in India. In S. B. Deo. (ed.). *Archaeological Congress Seminar Papers* Nagpur:
Vishnu-Mittre. 1974a. Palaeobotanical evidence in India. In J. Hutchinson (ed.). Evolutionary studies in world crops: Diversity and change in sub-continent. Cambridge: University Press pp. 3-30.
Vishnu-Mittre. 1974b. *Late Quaternary palaeobotany and palynology: an appraisement*. Lucknow: Birbal Sahni Institute of Palaeobotany.
Vishnu-Mittre. 1976. Comments on 'India', local and introduced crops by Sir Joseph Hutchinson. In The early history of agriculture. *Philosophical Transactions of the Royal Society of London* 275(936): 141.
Vishnu Mittre. 1978. Origins and history of agriculture in the Indian subcontinent. *Journal of Human Evolution* 7: 31-36.
Vishnu-Mittre, 1984. Quaternary palaeobotany/palynology in the Himalaya: an overview. *Palaeobotanist* 32(2): 158-187.
Vishnu-Mittre, 1985. The use of wild plants and the processes of domestication in the Indian subcontinent. In V. N. Misra and P. Bellwood (eds.). *Recent advances in Indo-Pacific prehistory*. New Delhi: Oxford and IBH Publishing Co.
Vishnu-Mittre and H. P. Gupta, 1968. Ancient plant economy at Paunar, Maharashtra. In S. B. Deo. *Report on the site*. Nagpur: Govt of India
Vishnu-Mittre and H. P. Gupta, 1968–69. Plant remains from ancient Bhatkuli, district Amraoti, Maharashtra. *Puratattva* 2:21-23.
Vishnu-Mittre, U. Prakash and N. Awasthi. 1971. Ancient plant economy at Ter, Maharashtra. *Geophytology* 1(2): 170-177.
Vishnu-Mittre and R. Savithri. 1974. Ancient plant economy at Noh. *Puratattva* 7:77-80.
Vishnu-Mittre and R. Savithri. 1975. Supposed remains of rice (*Oryza* sp.) in terracotta cakes and pai at Kalibangan, Rajasthan. *Palaeobotanist* 22(2):124-26.
Vishnu-Mittre and R. Savithri. 1978. Setaria spp. in the ancient plant economy of India. *Palaeobotanist* 25:559-562.
Vishnu-Mittre and R. Savithri. 1982. Food economy of Harappans. In L. Possehl (ed.). *Harappan civilization: a contemporary perspective*. New Delhi: Oxford & IBH, pp. 205-221.
Vishnu-Mittre and B. D. Sharma, 1966. Studies of post-glacial vegetational history from the Kashmir valley-I Haigam Lake *Palaeobotanist* 15c(1 and 2):85-212.
Wagnar, G. E. 1983. Late Harappan crops in Gujarat. *12th Annual Meetings Mid Atlantic Region of the Association for Asian Studies. University of Pennsylvania, Philadelphia*.
*Wang, S. C. 1984. *The Neolithic site of Chih-Shan-Yen*. Taipei, Taiwan: Literature Commission of Taipei City. 85pp.
Watkins, A.E. 1930. The Wheat species: a critique. *Genetics* 23: 173–263.
Watkins, R. 1976. Cherry, Plum, peach, apricot and almond. In N. W. Simmonds. *Evolution of crop plants*, London: Longman, 242-247.

Watson, W. 1969. Early cereal cultivation in China. In P. J. Ucko and G. W. Dimbleby. *The domestication and exploitation of plants and animals.* London: Duckworth.

Wendorf, F. R. Schild, N. Hadidi, El, A. E. Close, M. Kobusiewicz, H. Wieckowska, B. Issawi and H. Haas. 1979. Use of barley in Egyptian late Palaeolithic. *Science* 205:1341-1347.

Wendorf, F. and R. Schild. 1984. Some implications of the late Palaeolithic cereal exploitation at Wadi Kubbaniya Upper Egypt. In L. Krzyzaniak. and M. Kobusiewicz. (eds.). *Origin and early development of food producing cultures in Northeastern Africa* Poznen: Polish Academy of Sciences, pp. 117-127.

Whitaker, T. W. 1949. Identification and significance of the cucurbit materials from Huaca. *Prieta. Peru. Amer. Mus. Novet.* 1426.

Whitaker, T. W. 1981. Archaeological cucurbits. *Economic Botany* 35(4):460-466.

*Wittmack, L. 1880–1887. *Antike Samen aus Troja und Peru Monatsschr Ver Beford Gartenbaui.* Preussen:

Wolda, M. 1981. Similarity indices sample size and diversity. *Oecologia* 50:296-302.

Yacovleff, E. and F. L. Herrera. 1934–35. Elmundo vegetal de los antiques peruanos *Revista del Museo Nacional* 3:241-322; 4:29-102.

Yarnell, R.A. 1969. Palaeoethnobotany in America. In D. Brothwell and E. Higgs (eds.). *Science in Archaeology.* London: Thames and Hudson.

Yellen, J. E. 1977. *Archaeological approaches to the present: models for constructing the past.* New York: Academic Press.

Yen, D. E. 1982. Ban Chiang pottery and rice. *Expedetion* 24(4):51-64.

Zhimin, A. 1986. Prehistoric agriculture in China. *Recent Advances in the Understanding of Plant Domestication and Early Agriculture.* World Archaeological Congress Southampton, pp. 1-6.

Zohary, D. 1971. Origin of south-west Asiatic cereals, wheats, barleys oats and rye. In P. Davis et al. (eds.). *Plant life in Southwest Asia.* Edinburgh: University Press pp. 235-263.

Zohary, D. 1972. The wild progenitor and place of origin of the cultivated lentil: *Lens culinaris. Economic Botany* 26:326-332.

Zohary, D. 1976. Lentil. In N. W. Simmonds (ed.). *Evolution of crop plants.* London: Longman, pp. 163-164.

Zohary, D. 1986. The origin and early spread of agriculture in old world. In C. Barigozzi (ed.). *The origin and domestication of cultivated plants* Amsterdam: Elsevier, pp. 3-20.

Zohary, D. and M. Hopf. 1986. *Plants of Old World agriculture.* Oxford: University Press.

Zohary, D. and M. Hopf, 1973. Domestication of pulses in the Old World. *Science* 182:887-894.

Zohary, M. 1973. *Geobotanic foundations of the Near East.* 2 Vols. Stuttgart and Amsterdam: Springer.

Zukovskj, P. M. 1950. *Cultivated plants and their wild relatives.* Farnham, Bucks. Commonwealth Agricultural Bureau.

APPENDIX I
ECOLOGICAL CONSIDERATION OF PAST AND PRESENT VEGETATION OF KASHMIR VALLEY

Farooq A. Lone

INTRODUCTION

The valley of Kashmir (33° 30′–34° 30′ N; 74°–75° 30′ E), lying in an NE-SW direction, surrounded by the Himalayan range on the northeast and the Pir Panjal on the southwest, has a unique physical personality. Most of the valley is flanked by Carboniferous-Triassic rocks rising to altitudes of 3000 m and above. Rocks of Precambrian age are exposed around Baramulla and above the Banihal pass occur Jurassic rocks. The valley proper is filled by almost one kilometre deep sediments of a primeval lake, which was drained out later by emergence of the Jhelum river, forming what is known as 'Karewa series'. These relict lake sediments provide a continuous sediment record of climate and vegetation of the last four million years (Agrawal et al. 1985).

Kashmir valley is unique in respect of changes that have taken place in vegetation and climate in the late Cenozoic times. In the present work an attempt has been made to consider the past and present vegetation of Kashmir from an ecological point of view. The record of past vegetation has been obtained through palaeobotanic and palynological studies of the Karewa sediments. Now that the age of various localities in the Karewa series has been determined using magnetostratigraphy and fission track dating techniques (Burbank and Johnson 1982, Burbank 1985, Kusumgar et al. 1985), it has been possible to place various localities under geological

Ecological Consideration of Past and Present Vegetation of Kashmir Valley 257

epochs of Pliocene and Pleistocene. As many of the fossiliferous sites are yet to be accurately dated using modern physical dating techniques, they have been placed under Pleistocene based on relative lithostratigraphy. As has been well realized, it is difficult to analyse the data on past vegetation without accurate information on the present vegetation, a general survey of the present day vegetation of the valley has also been made.

The importance of micro- and megafossil evidence together in the reconstruction of past vegetation has been well emphasized as singular reliance on any one leads to erroneous conclusions (Vishnu-Mittre 1974). Therefore, as far as possible, the results of megafossil and pollen studies together have been used in inferring the past vegetation.

In the present attempt, in addition to our own data about the past and present vegetation of the valley, full use has been made of the work done by various workers on the palynology, palaeobotany and floristics of the valley as evidenced by the published work. These include Vishnu-Mittre (1965, 1984) and Puri et al. (1983) on palaeobotany, Vishnu-Mittre and Roberts (1971), Gupta et al. (1985) and Sharma et al. (1985) on palynology; Muthoo and Wali (1963), Wali (1964), Inayatullah and Tickoo (1964, 1965), Sapru et al. (1975), Singh and Kachroo (1976) and Singh and Kachroo (1983) on the floristics and vegetation.

To make the data on past and present vegetation more meaningful some statistical parameters have been applied. These include species diversity, species richness, species evenness and coefficient of similarity computed after Shannon and Weaver (1949), Margalef (1957) and Wolda (1981) as provided in Chapter 11. For all these analyses the data on arboreal vegetation alone have been used.

PLIOCENE VEGETATION OF KASHMIR

The evidence of the vegetation of Kashmir during Pliocene has come from the investigation at Dubjan and Hirpur localities in the Karewa Series falling in Gilbert and Gauss chrons in the magnetic polarity scale (Agrawal 1985).

The fossil woods recovered from Dubjan belong to *Abies pindrow, Pinus* sp and *Juglans* sp, Pollen analytical investigations have revealed *Pinus roxburghii, Abies, Picea, Cedrus, Cupressus, Taxus, Larix, Betula, Quercus, Corylus, Ulmus, Alnus, Juglans, Engelhardtia, Salix, Fraxinus, Viburnum, Acacia*, Fabaceae, Rosaceae, *Berberis*, Poaceae, *Artemisia*, Tubuliflorae, Liguliflorae, Chenopodiaceae, *Geranuim*, Urticaceae, *Polygonum*, Liliaceae, *Impatiens*, Cyperaceae, *Typha angustata, T. latifolia, Myriophyllum, Nymphaea, Potamogeton*, ferns and mosses. Statistical analysis of the data on arboreal vegetation revealed species diversity of 0.87, species richness of 3.94 and species evenness of 0.29.

The megafossil studies at Hirpur revealed *Pinus* sp, *Abies pindrow, Acer villosum, Aesculus indica, Fraxinus excelsior, Juglans regia, Quercus semicarpifolia, Q. dilatata, Populus* sp, *Viburnum aconitifolius, Salix wallichiana, Salix* sp, *Ulmus wallichiana, Typha angustata, Nelumbo* sp, *Potamogeton* sp and *Trapa* sp, Pollen analysis added *Pinus roxburghii, P. wallichiana, Picea, Cedrus, Juniperus, Cupressus, Larix, Betula, Corylus, Carpinus, Celtis, Alnus, Engelhardtia, Carya,* Fabaceae, Rosaceae, Oleaceae, Malvaceae, Poaceae, *Artemisia,* Tubuliflorae, Chenopodiaceae, Zygophyllaceae, *Geranium,* Scrophulariaceae, Caryophylaceae, Lamiaceae, Apiaceae, *Plantago,* Urticaceae, Ranunculaceae, *Polygonum,* Brassicaceae, *Impatiens,* Cyperaceae, *Myriophyllum, Eriocaulon, Lemna,* ferns, *Sphagnum,* mosses and algae like *Pediastrum,* and *Botryococcus.* The arboreal vegetation reveals a species diversity of 0.89, species richness of 4.26 and species evenness of 0.17.

Reconstruction of Pliocene Vegetation and plant communities around indicates that submerged plant community comprised of such species like *Potamogeton* and *Myriophyllum.* The free floating community included *Nelumbo* sp, *Nymphaea* sp, *Lemna* sp and *Trapa* sp, Among the reed swamps were growing *Typha angustata, T. latifolia* and some genera, of Cyperaceae. The shrubby vegetation included *Salix wallichiana, Salix* sp, *Berberis* sp, *Viburnum aconitifolius* etc. The forests comprised of conifers like *Pinus, Abies, Picea, Cedrus, Juniperus* and *Taxus.* The broad leaved elements included *Quercus semicarpifolia, Q. dilatata* and species of *Betula, Ulmus, Alnus, Acer, Juglans, Engelhardtia, Corylus, Carpinus* and *Populus.* Besides a large number of herbaceous species, ferns and mosses were also growing. The overall evidence indicates an essentially temperate climate.

PLEISTOCENE VEGETATION OF KASHMIR

The records of Pleistocene vegetation of Kashmir have come from sites like Laredura, Liddarmarg, Dangarpura, Nichahom, Ningle Nullah, Pakharpura, Khaigam, Aglar, Botapathri and Gogajipathri. The localities like Pakharpura, Khaigam and Aglar fall in the Matuyama and Brunhes chrons in the magnetic polarity stratigraphy. Rest of the sites have been placed under Pleistecene as their magnetic stratigraphy has not been worked out so far.

The plant fossils from Laredura belong to *Mallotus philippinensis, Aesculus indica, Myrsine* sp, *Olea grandulifera, Fraxinus* sp, *Ulmus* sp, *Engelhardtia colebrookiana, Woodfordia fruticansa, Salix elegans, Salix* sp, *Quercus semicarpifolia, Q. dilatata, Q. ilex, Betula utilis, B. alnoides, Alnus nitida, Corylus ferox, Acer villosum, A. caesium, A.* sp, *Berberis lycium, Hedera nepalensis, Desmodium nutans, D. latifolium, D. tiliaefolium, Indigofera hebepetala, I.* sp, *Rhus punjabensis, R. succedania, Odina wodier, Prunus cerasoides, P.* sp, *Pyrus pashia, Rosa webbiana, R. macrophylla, R.* sp, *Spiraea* sp, *Cotoneaster bacillaris, Rubus fruticosus, R.* sp, *Trapa natans, T. bispinosa, Ceratophyllum.*

sp, *Myriophyllum* sp, *Nymphaea* sp and *Euryale ferox*. The pollen belong to *Pinus wallichiana, Cedrus, Picea, Abies, Ephedra, Quercus, Ulmus, Alnus, Carpinus, Juglans, Aesculus, Acer, Rhus, Fraxinus, Populus, Salix, Corylus, Viburnum*, Polypodiaceae, Rosaceae, Polygonaceae, Ranunculaceae, Umbelliferae, Caryophyllaceae, *Impatiens*, Chenopodiaceae, Compositae, Poaceae, Cyperaceae, *Typha, Trapa, Nymphaea, Lemna* and *Potamogeton*. The arboreal vegetation shows species diversity of 1.59, species richness of 6.78 and species evenness of 0.43.

The megafossils from Liddarmarg belong to *Quercus incana, Q. glauca, Q.* sp, *Ficus cunia, Mallotus phillipensis, Acer oblongum, A. pentapomicum, A.* sp, *Litsaea lanuginosa, Machilus odoratissima, M. duthie, Phoebe lanceolata, Buxus wallichiana, B. papillosa, Ulmus* sp, *Skimmia laureola, Toddalia* sp, *Pittosporum eriocarpum, Rhamanus virgatus, R. triquetra, Berchemia floribunda, Myrsine africana, M. semiserrata, Syringe emodi, Wendlandia exserta, Pyrus communis, Cotoneaster bacillaris, Spiraea* sp, *Betula utilis, Alnus nepalensis, Berberis lycium, Berberis* sp, *Dendrobenthamia capitata, Parrotiopsis jacquemontiana, Desmodium podocarpum, D. laxiflorum, Inula cappa, Acorus* sp, *Scirpus* sp, *Cyperus* sp, and Cyperaceae. The palynology of the sediments revealed *Pinus wallichiana, Cedrus, Picea, Abies, Ephedra, Quercus, Betula, Aesculus, Acer, Rhus, Populus, Salix, Corylus, Parrotiopsis, Viburnum*, Polypodiaceae, Rosaceae, Ranuculaceae, Umbelliferae, Caryophyllaceae, *Impatiens*, Chenopodiaceae, Poaceae, Cyperaceae, *Typha*, and *Botryococcus*. The arboreal vegetation shows species diversity of 1.19, species richness of 8.68 and species evenness of 0.32.

From Dangarpura the fossils belong to *Quercus semicarpifolia, Q. dilata, Q. ilex, Hedera nepalensis, Litsaea elongata, L.* sp, *Trapa natans, T. bispinosa, Typha*, sp *Sparganium* sp, *Ceratophyllum* sp, *Myriophyllum* sp, *Scirpus* sp, and *Cyperus* sp. Pollen analysis revealed *Pinus wallichiana, Cedrus, Picea, Abies, Ephedra, Quercus, Ulmus, Juglans, Aesculus, Acer, Salix, Corylus, Parrotiopsis,* Polypodiaceae, Rosaceae, Polygonaceae, Ranunculaceae, Chenopodiaceae, Asteraceae, Poaceae, Cyperaceae, *Typha, Lemna* and *Botryococcus*, Aboreal vegetation shows species diversity of 0.98, species richness of 3.90 and species evenness of 0.34.

At Nichahom the megafossils belong to *Acer caesium, A. pentapomicum, A. pictum, A.* spp, *Aesculus indica, Alnus* sp, *Berberis* sp, *Buxus papillosa, Cotoneaster bacillaris, C. microphylla, Hedera helix, Pittosporum eriocarpum, Pyrus pashia, Quercus dilatata, Q. glauca, Q. incana, Q. ilex, Q. semicarpifolia, Ulmus compestris*, few leaves of grasses and ferns. Pollen analysis revealed *Pinus wallichiana, Cedrus, Picea, Abies, Quercus, Ulmus, Alnus, Juglans, Aesculus, Acer, Rhus, Fraxinus, Salix, Corylus, Viburnum*, Polypodiaceae, Liliaceae, Caryophyllaceae, *Impatiens*, Chenopodiaceae, *Artemisia*, Poaceae, Cyperaceae, *Typha, Trapa, Myriophyllum, Potamogeton* and *Botryococcus*. Arboreal vegetation reveals species diversity of 1.06, species richness of 3.9 and species evenness of 0.36.

At Ningle Nullah the fossils recovered are those of *Aesculus indica, Ulmus laevigata, Salix wallichiana, Salix elegans, Salix* sp, *Populus ciliata, P. balsamifera, Populus* sp, *Betula utilis, Alnus nepalensis, Acer pentapomicum, A. villosum, A. pictum, A.* sp, *Cornus macrophylla, Marlea begoniaefolia, Inula cappa, Desmodium gangeticum, Nelumbo nuncifer, Fraxinus excelsior, Typha* sp, *Sparganium* sp, *Prunus cornuta, Prunus* sp, *Cotoneaster nummularia* and *C. microphylla.* Pollen analysis of these sediments revealed *Pinus wallichiana,*—*Cedrus, Picea, Abies, Ephedra, Quercus, Betula, Ulmus, Alnus, Juglans, Aesculus, Fraxinus, Populus, Engelhardtia, Salix, Corylus,* Polypodiaceae, Ranunculaceae, Apiaceae, Caryophyllaceae, *Impatiens,* Chenopodiaceae, Poaceae, Cyperaceae, *Typha, Lemna, Potamogeton* and *Botryococcus.* Species diversity of 1.46, species richness of 7.38 and species evenness of 0.41 is revealed by the data on arboreal vegetation.

At Pakharpura the megafossils belong to *Aesculus indica Populus* sp, *Juglans regia, Salix alba, Rubus ulmifolius, Ulmus wallichiana, Quercus* sp, *Acer* sp, *Typha angustata, Trapa natans, T. bispinosa* and some borage. Pollen analysis of these sediments is yet to be carried out. The megafossil evidence on arboreal vegetation reveals species diversity of 0.82, species richness of 2.91 and species evenness of 0.39.

At Khaigam the fossil woods recovered belong to *Pinus* sp, *Pinus wallichiana, Picea smithiana* and *Populus* sp, Aglar revealed scores of leaf impressions belonging to *Ulmus* sp and *Salix* sp, Besides, *Nymphaea* sp, *Parrotiopsis jacquemontiana, Typha* sp and *Trapa* sp were also recovered. From Gogajipathri *Ulmus compestris, Quercus semicarpifolia, Rhus cotinus* and *Sparganium* sp and from Botapathri *Quercus incana, Q. glauca* and *Trapa natans* are known. Pollen analysis of these sediments is yet to be carried out.

Reconstructing the plant communities growing around during Pleistocene it becomes evident that the submerged plant community in the Karewa lake was comprised largely by the species of *Ceratophyllum* and *Myriophyllum.* Free floating plant community included species of *Trapa, Lemna* and *Nelumbo* and reed swamps were constituted of species of *Typha, Sparganium, Cyperus, Scirpus* and *Acorus.*

Conifers occurred in the forests. Very common were *Pinus wallichiana* and *Abies.* Others included *Picea, Cedrus, Juniperus, Ephedra* and *Taxus.* Amongst the broad leaved trees the oaks dominated and were represented by many species. The temperate deciduous genera such as *Acer, Carpinus, Aesculus, Prunus, Ulmus* and *Betula* were of fair distribution. *Litsaea, Machilus* and *Cinamomum* were present. The other constituents included *Fraxinus, Rhamnus, Meliosma, Euonymus, Ilex,* birches and alders. The shrubby vegetation comprised of *Rosa macrophylla, Berberis aristata, Salix elegans* and species of *Rubus, Spiraea, Viburnum, Desmodium* and *Indigofera.*

Some species of *Rosa*, *Hedera* and *Clematis* comprised the climbers. Herbaceous species belonged to many genera of a large number of families. Some ferns and mosses were also growing.

The tropical elements in the Pleistocene flora included *Ficus cunia, Mallotus philipensis, Pittosporum eriocarpum, Wendlandia exserta, Berchemia, Myrsine, Odina wodier, Litsaea lanuginosa, Woodfordia fruticosa, Buxus* spp and *Machilus* sp, Thus during Pleistocene in addition to essentially temperate genera some sub-tropical and tropical plants were also growing in the valley.

PRESENT VEGETATION OF KASHMIR

On the basis of extant records 2946 species belonging to 741 genera are represented in the valley (Kachroo 1986). A general survey of arboreal vegetation from near 1350 m to the upper altitudinal level of the plant life reveals following vegetational zones.

Zone I: The Valley Proper Zone

This zone is occupied by the agricultural fields, orchards, water bodies, pastures, habitations and graveyards etc. The common broad leaved tree elements growing in this zone comprise usually the planted species like *Salix wallichiana, Populus nigra* var. *fastigiata, P. alba, P. ciliata, Aesculus indica, Ulmus wallichiana, Celtis australis, Platanus orientalis, Fraxinus excelsior, Juglans regia, Robinia pseudoacacia, Morus alba, M. nigra, Ailanthus* sp, *Malus sylvestris, Prunus amygdalus, P. persica, P. armeniaca, P. cerasifera, P. domestica* etc. Sometimes trees of *Zizyphus jujuba, Quercus robur, Cupressus* sp and *Pinus wallichiana* are also met with.

Zone II: Lower Conifer Zone

This zone extends from 1500m to c.2100m. The conifers *Pinus wallichiana* and *Cedrus deodara* dominate the vegetation. Also recorded in this zone are *Parrotiopsis jaequemontiana, Viburnum nervosum, Berberis* sp, *Acer caesuim, Alnus nitida, Rhus* sp, *Morus* sp, *Prunus* spp, *Fraxinus* sp, *Celtis* sp, *Indigofera* sp, *Juglans* sp, *Robinia pseudoacacia* and *Quercus robur*. The conifer species are sometimes seen forming pure patches and sometimes form associations with the broad leaved elements. The common associations found in this zone are:

(a) *Cedrus deodara-Parrotiopsis* association.
(b) *Pinus wallichiana-Cedrus deodara* association.
(c) *Cedrus deodara-Pinus wallichiana-Parrotiopsis- Viburnum-Indigofera* association.

(d) *Pinus wallichiana-Parrotiopsis* association and
(e) *Pinus-Viburnum* association.

Zone III: Upper Conifer Zone

This zone extends from c. 2000 m to 3000 m. The plant species met with in this zone include *Abies pindrow, Picea smithiana, Taxus wallichiana, Padus cornuta, Aesculus indica, Acer caesium, Juglans regia, Fraxinus hookeri, Crataegus oxyacantha, Rosa macrophylla, Ribes* sp, *Lonicera* sp, *Viburnum nervosum, Parrotiopsis jacquemontiana, Quercus semicarpifolia* and *Q. dilatata.* Common associations found are:

(a) *Abies-Acer-Aesculus,-Juglans-Viburnum-Parrotiopsis* association.
(b) *Abies-Pinus-Parrotiopsis-Viburnum* association
(c) Pure *Abies* community
(d) *Abies-Pinus*—broad leaf/shrub association
(e) *Abies-Picea-Pinus* association
(f) *Abies-Picea*—broad leaf shrub association
(g) *Abies*-mixed evergreen broad leaf shrub association.
(h) *Abies-Pinus* association.

Zone IV: White Birch Zone

This zone extends from 3100 m to 3600 m and is characterized by the presence of white birch *Betula utilis* which forms pure monotypic stands at certain places. Other tree species recorded in this zone are *Rhododendron anthopogon, R. companulatum, Salix denticulata, Syringe emodi* and *Lonicera* sp, Also recorded at some places are *Abies pindrow, Quercus* spp, *Juglans* spp, *Acer* sp, and *Populus* spp. The common associations found in this zone are:

(a) *Betula utilis-Rhododendron anthopogon* association
(b) *Betula utilis-Rhododendron companulatum* association
(c) *Betula utilis-Abies pindrow* association

Zone V: The Alpine Zone

This zone extends from 3600 m–4000 m and lies well above the upper limit of conifer zone. It supports a thick vegetation of *Juniperus communis* and *J. recurva.* Also recorded in this zone are some willows, *Lonicera* sp, *Rhododendron anthopogon, R. companulatum, Syringe emodi, Cotoneaster numularia, C. racemofolia, Gaultheria mummularioides* and *Prunus prostrata.* The main plant associations occurring in this zone are:

(a) *Juniperus recurva-Rhododendron anthopogon* association

(b) *Juniperus communis-R. companulatum* association
(c) *Rhododendron companulatum-Syringe emodi* association
(d) *Juniperus communis-Rhododendron anthopogon* association.

As far the present-day plant communities are concerned the submerged plant community comprises of *Potamogeton, Hydrilla, Myriophyllum, Ceratophyllum* and *Chara*. The free floating community is constituted by species of *Nymphaea, Nelumbo, Lemna, Euryale, Trapa* and *Salvinia*. Reed swamps comprise of species of *Typha, Sparganium, Scirpus, Carex, Acorus, Callitriche, Juncus, Marsilia* and *Utricularia*.

The shrubby vegetation consists of species of *Salix, Rubus, Rosa, Viburnum, Parrotiopsis, Morus, Crataegus, Astragalus, Indigofera, Ribes* etc. The common conifers are *Pinus wallichiana, Cedrus deodara, Picea smithiana* and *Abies pindrow*. The common broad leaved trees are *Populus, Ulmus, Acer, Fraxinus, Juglans* and *Aesculus*. The climate on the whole falls under temperate.

STATISTICAL OBSERVATIONS

The aforementioned account gives a clear picture of the vegetation of Kashmir during Pliocene, Pleistocene and present times. In order to get an insight into the relationships of the vegetation, coefficient of similarity has been computed between pairs of sites as provided in table 1. Various fossiliferous sites and the present-day vegetation zones have been involved in this analysis. The values reveal the extent of likness or unlikeness between the pairs. The values of 'S' vary from 1.00, when all the species are common, to 0.0, when none of the species are common. The values in between indicate the extent of similarity.

Index of similarity was also computed for the Pliocene, Pleistocene and present-day vegetation as a whole (table 2). Values of 0.61 between Pliocene and Pleistocene, 0.56 between Pliocene and present and 0.44 between Pleistocene and present were obtained. The evidence available at present indicates that the vegetation has not be alike throughout.

The values of species diversity, species richness and species evenness have been calculated for each site and are quite significant. Species diversity is a measure that takes into account both the total number of species or taxa present in a population and the abundance of each (Pielou 1969). High diversity results when a large number of species are evenly distributed i.e. when it would be difficult to predict what a randomly selected item would be. Low diversity results when the number of species is variable. General diversity is affected by both the number of species present and the evenness with which they are distributed (Potter and Kesselle 1980). To separate these effects Margalef's (1957) measure of species richness and Shannon and Weaver's (1949) index of species evenness were used in the present

Table 1. Index of similarity(s) including arboreal vegetation of various fossiliferous sites and present-day vegetation Zones.

	Dubjan	Hirpur	Laredura	Liddarmarg	Dangarpur	Nichahom	Ningle Nullah	Pakharpura	Kashmir Valley Zone	Lower Conifer Zone	Upper Conifer Zone	White Birch Zone	Alpine Zone
Dubjan	1.00	0.58	0.41	0.38	0.50	0.48	0.48	0.32	0.27	0.39	0.40	0.32	0.0
Hirpur			0.48	0.54	0.57	0.49	0.49	0.47	0.31	0.37	0.42	0.38	0.05
Laredura				0.45	0.58	0.58	0.43	0.35	0.23	0.32	0.30	0.24	0.04
Liddar marg					0.46	0.56	0.41	0.32	0.23	0.29	0.29	0.30	0.0
Dangarpur						0.54	0.35	0.55	0.33	0.48	0.48	0.34	0.0
Nichahom							0.61	0.37	0.21	0.33	0.34	0.23	0.0
Ningle Nullah								0.38	0.23	0.27	0.31	0.26	0.0
Pakhar Pura									0.37	0.21	0.29	0.33	0.0
Kashmir valley zone										0.28	0.19	0.16	0.0
Lower conifer zone											0.28	0.14	0.0
Upper conifer zone												0.29	0.0
White Brich zone													0.15
Alpine zone													1.00

study. The respective values for different sites are indicative of the extent of stability of vegetational populations growing around.

Table 2. Index of similarity involving arboreal vegetation of Kashmir during Pliocene, Pleistocene and present times.

	Pliocene	Pleistocene	Present
Pliocene	1.00	0.61	0.56
Pleistocene		1.00	0.44
Present			1.00

CONCLUDING REMARKS

Over viewing the entire mass of the botanical information, it is to be admitted that this botanical parameter hardly shares any iota of credit for delimitation of Pliocene-Pleistocene boundary. The botanical evidence makes us to believe that the Pliocene floristics were very much like the early Pleistocene floristics and there was hardly anything like the transition from one into another. However, it appears that during Pliocene the temperate climate and vegetation dominated the valley whereas during Pleistocene many sub-tropical and tropical elements were also growing which would mean some change in climatic conditions. On the other hand the credit in this potential area goes to palaeomagnetic and fission track dating, however insufficent (Agrawal et al. 1981, Burbank and Johnson, 1982, Burbank 1985, Kusumgar et al. 1985). More work is indeed needed in this direction particularly for the localities from where the fossils have been reported.

Most of the reconstructed evidence on past vegetation is indicative of the prevalence of a temperate climate. However, as already pointed out, some tropical/subtropical elements also occurred. This has led some workers (Puri et al. 1983) to conclude that during Pleistocene times tropical climate prevailed in the valley. However Vishnu-Mittre (1984) points out that these forms today extend up to the threshold of temperate belt. None of them is represented in the present-day flora of the valley.

There are some other elements also which occur in the Pliocene/Pleistocene deposits but are unrepresented in the modern flora of the valley and of these *Quercus* spp. is the most important. Many species are represented in the Pleistocene flora like *Q. semicarpifolia*, *Q. dilatata*, *Q. glauca*, *Q. incana*, *Q. ilex*. These are absent from the valley present except for a few occurrences of *Q. dilatata* and *Q. semicarpifolia* (Vishnu-Mittre 1963). According to Puri et al. (1983) the present conditions are brought about by the lofty ranges of the Pir Panjal which forms a formidable barrier to the monsoon winds. The Himalayas were, indeed, much lower than now during Pleistocene and the uplift of Pir Panjal is responsible for the destruction of tropical rain forest elements by denying them the heavy rain fall. However, *Quercus* shows fair to good frequencies in the post glacial pollen diagrams from the valley (Singh 1963, Vishnu-Mittre and Sharma 1966) and has also been reported from the charcoals of archaeological excavations (Lone 1987). Therefore, it appears that oaks have not been omitted in the Pleistocene but have been growing over here up to much later times. In addition to climatic factors, human hand appears to be involved in their reduction.

Vishnu-Mittre (1984) has pointed out that a fact of considerable significance is the discovery at the same site and altitude of fossils of ecologically incompatible taxa distributed today at different altitudes in the Himalaya, for example of *Betula utilis, Quercus glauca, Q. incana* at Liddarmarg; *Betula utilis* and *Alnus nepalensis* at Ningle Nullah, *Quercus semicarpifolia* and *Litsea elongata* at Dangarpura and *Q. semicarpifolia, Q. incana* and *Q. dilatata* at *Laredura*. This leads one to believe that climatic requirements of these taxa must have been different during the early Pleistocene. However, it is most likely that the leaves of these species have been brought from various altitudes through the agencies of winds, water etc. and deposited in the lake sediments.

Lastly I would like to comment upon the identification of megafossils and Quaternary pollen from Kashmir. Some doubts have been raised about the identification of some megafossil leaves and pollen of '*Larix*' and '*Pinus roxburghii*' (Vishnu-Mittre 1984). Therefore, it is timely to re-examine these evidences so that inferences on past vegetation are not affected.

REFERENCES

Agrawal, D.P. (1985). Cenozoic climatic changes in Kashmir: the multidisciplinary data. *Current Trends in Geology Vol. VI (Climate and Geology of Kashmir)* New Delhi: Today and Tomorrow Printers and Publishers, pp 1-12.

Agrawal, D.P., R.N. Athavale, R.V. Krishnamurthy, S. Kusumgar, C.R.K. Murthy, and V. Nautiyal. (1979). Chronostratigraphy of loessic and lacustrine sediments in Kashmir valley. *Acta. Geol. Acad Scient Hungaricae Tomus* 22(1-4):185-196.

Agrawal, D.P., S. Kusumgar, and R.V. Krishnamurthy. (1985). *Climate and Geology of Kashmir and Central Asia: The last four Million years.* New Delhi: Today & Tomorrow's Printers and Publishers.

Burbank, D.W. (1985). The age of Karewas, Kashmir, as determined from fission track dating and magnetostratigraphies. *Current Trends in Geology.* Vol. VI (Climate and Geology of Kashmir), New Delhi: Today and Tomorrow printers and Publishers, pp. 19-26.

Burbank, D.W. and Johnson, G.D. (1982). Intermontane basin development in the past 4 Myr in the northwest Himalaya. *Nature* 298:432-436.

Gupta, H.P., C. Sharma, R. Dodia, C. Mandavia, and A.B. Vora. (1985). Palynostratigraphy and Paleoenvironments of Kashmir: Hirpur locality III. *Current Trends in Geology* Vol. VI *(Climate and Geology of Kashmir)*: New Delhi: Today and Tomorrow Printers and Publishers, pp. 75-90.

Inayatullah, M. and B.L. Ticku. (1964). A preliminary study of the forest topology of J & K. *Indian Forester* 90(6) : 332-341.

Inayatullah, M. and B.L. Ticku. (1965). Ecological study of the forest types in Lolab valley and adjoining areas. *Indian Forester* 91(8) : 538-547.

Kachroo, P. (1986). A note on flora of Kashmir. In M. Hussain et al. *Geography of Jammu and Kashmir (some aspects)*. New Delhi: Ariana Publishers, pp. 32-44.

Kusumgar, S., N. Bhandari, and Agrawal D.P. (1985). Fission track ages of the Romushi, Lower Karewa, Kashmir. *Current Trends in Geology* (Climate and Geology of Kashmir). New Delhi: Today and Tomorrow Printers and Publishers, pp. 245-247.

Lone, F.A. (1987). *Palaeoethnobotanical Studies of Archaeological Sites of Kashmir*. Unpublished Ph. D. thesis, University of Kashmir.

Margalef, R. (1957). Information theory in ecology. *General Systems Bulletin* 31:36-71.

Muthoo, M.K. and M.K. Wali. (1963). Deodar belt of Kashmir, Lolab valley. *Indian Forester* 89 :716.725.

Pielou, E.C. (1969). *An introduction to Mathematical Ecology*, New York.

Potter, M.W. and S.P. Kessell. 1980. Predicting mosaics and wildlife diversity resulting from fire disturbance to the forest ecosystem. *Environmental Management* 4: 247-254.

Puri, G.S. Meher, V.M. Homji, R.K. Gupta and S. Puri. (1983). *Forest Ecology* Vol. I, New Delhi: Oxford and IBH Publishing Co.

Sparu, B.L., U. Dhar and P. Kachroo, (1975). Vegetation studies in Jhelum valley. *Botanique* 6:151-164.

Shannon, C.E. and N. Weaver. (1949). *The Mathematical Theory of Communication*. Urbana: University of Illinois Press.

Sharma, C., H.P. Gupta, R. Dodia and C. Mandavia. (1985). Palynostratigraphy and palaeoenvironments: Dubjan, Lower Karewa, Kashmir. *Current Trends in Geology* Vol. VI (Climate and Geology of Kashmir). New Delhi: Today and Tomorrow Printers and Publishers, pp. 69-74.

Singh, G. (1963). A preliminary survey of the post-glacial vegetational history of Kashmir valley. *Palaeobotanist*: 12(1): 73-108.

Singh, G. and P. Kachroo. (1976). *Flora of Srinagar and plants of neighbourhood*. Dehra Dun: Bishen Singh Mahendra Pal Singh.

Singh, J.B. and P. Kachroo. (1983). Plant community characteristics in Pir Panjal Forest range (Kashmir). *Journal Economic Taxonomic Botany* 4(3):911-937.

Vishnu-Mittre, G. Singh and K.M.S. Saxena. 1962. Pollen analytical investigations of lower Karewa. *Palaeobotanist* 11:92-95.

Vishnu-Mittre (1965). Floristic and ecological reconsiderations of the Pleistocene plant impressions from Kashmir. *Palaeobotanist* 13(3):308-327.

Vishnu-Mittre (1974). Quaternary vegetation in Northern region. In Surange K.R. et al. (eds.) *Aspects and Appraisal of Indian Palaeobotany*, Lucknow: B.S.I.P., pp. 657-664.

Vishnu-Mittre (1984). Quaternary palaeobotany/palynology in the Himalaya: an overview. *Palaeobotanist* 32(2) : 158-187.

Vishnu-Mittre and R. Robert. (1971). Pollen analysis and palaeobotany of impression bearing sediments in the Lower Karewas. *Palaeobotanist* 20(3):344-355.

Vishnu-Mittre and B.D. Sharma. (1966). Studies of post-glacial vegetational history from the Kashmir valley-I. Haigam lake. *Palaeobotanist* 15(1, 2): 185-212.

Wali, M.K. (1964). A preliminary survey of the conifer communities of Kashmir, Himalayas. *Tropical Ecology* 5:32-41.

Wolda, M. (1981). Similarity indices, sample size and diversity. *Oecologia* 50: 296-302.

APPENDIX II

KASHMIR—ETHNOBOTANIC PRESENT

Maqsooda & Farooq A. Lone

In the following lines we present a check list of plant species commonly used in Kashmir valley as food plants, vegetables, pulses, spices & condiments, narcotics & beverages, oilseeds, fruits, fibres and medicinal plants so as to give the reader a general idea about the present-day ethnobotany of the valley. Botanical names are followed by local Kashmiri names in the parenthesis.

A. FOOD PLANTS

a) MAJOR: *Oryza sativa* (Dhani, Tomool); *Tritium aestivum* (Kanik); *Zea mays* (makai); *Hordeum vulgare* (Vishka).

b) MINOR: *Panicum miliaceum* (Pinga); *Setaria italica* (Shola); *Fagopyrum esculentum* (Tromba); *Saccharum officinarum* (Khand); *Amaranthus gangeticus* (Ganhar); *Nymphaea candida* (Ken Buch); *Carduus nutans* (Chari Tomool); *Meteroxylon sago* (Sabudan).

B. PULSES

Phaseolus aureus (Mong); *P. mungo* (Mah); *P. aconitifolius* (Muth); *P. vulgaris* (Razmah); *Lens culinaris* (Masur); *Pisum sativum* (Kar); *Vicia faba* (Bhogla); *Cicer arietinum* (Chana, Chola); *Glycine max* (Soya bean); *Lathyrus sativus* (Neuk Kar).

C. VEGETABLES

Solanum tuberosum (Aalu); *S. melanogena* (Wangun); *Rumex nepalensis* (Abuj); *Cucurbita pepo* (Al); *C. maxima* (Al); *Brassica oleracea var capitata* (Band Gobi); *B. oleracea vari botrytis* (Phool Gobi; *B. oleracea* var *caulorapa* (Haak); *Podophyllum emodi* (Ban Wangun); *Hibiscus esculentus* (Bhindi); *Beta vulgaris* (Chukander); *Rumex acetosa* (Tsuk-Tsen); *Centuria iberica* (Kurtsch); *Allium cepa* (Ganda); *Daucus carota* (Gazar); *Brassica rapa* (Gogaj); *Morchella* esculenta (Guch); *Taraxacum officinale* (Hand); *Rumex orientalis* (Jungli Abuj); *Momordica charantia* (Karela); *Amaranthus caudatus* (Lisa); *Trigonella foenumgracum* (Meeth); *Raphanus sativus* (Muj); *Nelumbo nucifera* (Nadur); *Phytolacca acinosa* (Nunar); *Portulaca oleracea* (Nunar); *Spinacea oleracea* (Palak); *Lycopersicum esculentum* (Ruwangun); *Malva rotundifolia* (Sochal); *Luffa acutangula* (Tareela); *Polygonum persicaria* (Tsokaladder); *Chenopodium foliosum* (Wan Palak); *Chenopodium* sp (Wasta Haak); *Dipsacus intermis* (Wapal Haak); *Brassica compestris* (Til Gugal Hak); *Cucumis sativus* (Ler); *Psalliota compestris* (Hedur).

D. SPICES AND CONDIMENTS

Brassica oleracea (Aasur); *Foeniculum vulgare* (Badiana); *Coriandrum sativum* (Dhan Wal); *Piper nigrum* (March); *Capsicum frutescens* (Marcha Wangun); *Celosia cristata* and *C. argentea* (Ma Wal); *Allium cepa* var *vivivora* (Pran); *A. sativum* (Rohan); *Cuminum cyminum* (Ziur); *Ferula northex* (Yang); *Amomum aromaticum* (Buda Ael); *Cinnamomum zeylanicum* (Dalchin); *Curcuma longa* (Ledher); *Elletaria cardamomum* (Sabz Ael); *Eugenia aromatica* (Roung); *Zingiber officinale* (Shaunt); *Tamarix indicus* (Tamber); *Crocus sativus* (Kaung)

E. NARCOTICS AND BEVERAGES

Papaver somniferum (Afeen); *Cannabis sativus* (Bhang, Charas); *Nicotiana tabacum* (Tamouk); *Camellia sinensis* syn *Thea* sinensis (Chai); *Coffea arabica* (Kafi); *Piper betle* (Pan); *Areca catechu* (Supari).

F. OILSEEDS

Brassica compestris (Til Gogul); *Linum usitatissimum* (Alish); *Juglans regia* (Doon); *Cocos nucifera* (Khopra, Narfeel); *Sesamum indicum* (Tel); *Helianthus* annus (Gul-e-Aftab); *Ricinus communis* (Cash Tur), *Olea europea* (Zytoon)

G. FRUITS

Prunus cerasus (Aer); *P. domestica* (Alich); *P. Persica* (Tsunun); *P. amygdalus* (Badam); *P. armeniaca* (Tser); *P. avium* (Gilas); *P. bokhariensis* (Aalu-Bukhara); *Ficus carica* (Anjoor); *Phyllanthus emblica* (Amla); *Morus alba* (Bota Tul); *M. rubra* (Shah Tul); *Rubus lasiocarpus* (Chaunch);

Malus sylvestris (Tsunt); *Vitis vinifera* (Dachh); *Punica granata* (Daen); *Juglans regia* (Doon); *Pyrus communis* (Tang); *P. pashia* (Taenj); *Trapa bispinosa* (Gor); *Citrullus vulgaris* (Hend Wend); *Fragaria vesca* (Ishtabur); *Cucumis melo* (Kharbuza); *Nelumbium speciosum* (Pambach); *Psidium guyava* (Amrood); *Ananas comosus* (Pineapple); *Anacardium occidentale* (Kaju); *Musa sapientum* (Kela); *Phoenix dactylifera* (Khazir); *Cocos nucifera* (Khupra); *Citrus* spp (Sangtara, Mausmi, Leum); *Carica pappaya* (Papeeta); *mangifera indica* (Amb).

H. FIBRE AND MATTING

Gossypium hirsutum (Sether); *Cocos nucifera* (Tat); *Corchorus* sp (Suthal); *Oryza sativa* (Gass Pataj); *Typha angustata* (Wagu); *Ulmus wallichiana* (Bren Del); *Cannabis sativa* (Bhang-e-Del).

I. RELIGIOUS CEREMONIES AND RITUALS

Santalum album (Tsandun); *Lawsonia inermis* (Maenz); *Aconitum heterophyllum* (Nar-Mad); *Iris nepalensis* (Mazar Mond); *Peganum harmala* (Isband); *Jurinea longifolia* (Dhupa); *Ferula northex* (Yang); *Celtis australis* (Bremij); *Oryza sativa* (Tamul); *Triticum aestivum* (Kanik); *Hordeum vulgare* (Vishka); *Sesamum indicum* (Tel); *Juglans regia* (Doon); *Saccharum officinarum* (Nabad, Kand); *Saussurea sacra* (Jugi Padsha).

J. MEDICINAL PLANTS

Calendula arvensis (Bat Posh); *Rumex nepalensis* (Abuj); *Ficus cariaca* (Anjoor); *Foeniculum vulgare* (Badiana); *Cotula anthemoides* (Bobal); *Ocimum sanctum* & *Leonurus cardiaca* (Brari Gass); *Zizyphus Jujuba* (Brai); *Salix capria* (Bred Mushk); *Cydonia oblonga* (Bum Tsunt); *Viola odorata* (Bunafsha); *Descurina sophia* (Chari Lasehij); *Sisymbrium loeselii* (Dand-Haak); *Coriandrum sativum* (Dhanwal); (*Datura stramonium*) (Datur); *Cassia fistula* (Faloose); *Nepeta cataria* (Gandh Soi); *Adiantum capillus veneris* (Gew Theer); *Plantago lanceolata* (Gula); *P. ovata* (Ismagool); *Chrysanthemum morifolium* (Gul-e-Dawood); *Plumeria alba* (Gul Cheen); *Taraxacum officinale* (Hand); *Cichorium intybus* (Handi Posh); *Saussurea lappa* (Kuth) *Thymus serphyllum* (Javend); *Macrotomia benthami* (Kah Zaban); *Prunella vulgaris* (Kal Yuth); *Glycerhiza glabra* (Shanger); *Dioscoria deltoides* (Kiddri); *Berberis lycium* (Kaw Dach); *Cinnamomum camphora* (Kafoor); *Crocus sativus* (Kung); *Pteris vittata* (Loosa Gass); *Tribulus terrestris* (Mecher Kund); *Celosia* spp (Mawal); *Gentiana* kurroo (Nilkanth); *Rubus* spp. (Pamb Chalan); *Mentha arvensis* (Pudina); *Allium sativum* (Ruhan); *Althaea rosea* (Saz mool); *Fumaria* spp (Shah Tar); *Artemisia absinthium* (Teth Wen); *Co-*

tula anthemoides (Thul Bobul); *Marrubium* sp. (Troper); *Sisymbrium brassiforme* (Tseri Lachif); *Mentha* spp (Veun); *Digitalis purpurea* (Digitalis); *Cordia* spp (Sapistan); *Rosa* spp (Gulab); *Lawsonia inermis* (Maenz); *Smilax china* (Chob cheen).

APPENDIX III
SOFT AND HARD WOODS OF KASHMIR

Farooq A. Lone & Maqsooda

A large number of soft and hard woods are used as timber and fuel wood in Kashmir. The important woody plant species constituting the arboreal vegetation of the valley are as follows:

SOFT WOODS

1. *Abies pindrow* 2. *Cedrus deodara* 3. *Cupressus sempervirens* 4. *Biota orientalis* 5. *Picea* smithiana 6. *Pinus wallichiana* 7. *Juniperus communis* 8. *Taxus baccata*

HARD WOODS

1. *Populus alba* 2. *P. nigra* 3. *P. euphratica* 4. *P. deltoides* 5. *P. ciliata* 6. *Salix wallichiana* 7. *S. alba* 8. *salix* spp. 9. *Juglans regia* 10. *J. nigra* 11. *Betula utilis* 12. *Alnus nitida* 13. *Corylus Jacquemontii* 14. *C. Colurna* 15. *Carpinus* spp. 16. *Castanea sativa* 17. *Quercus robur* 18. *Q.* semicarpifolia 19. *Celtis australis* 20. *Ulmus wallichiana* 21. *U. villosa* 22. *Ficus* spp. 23. *Morus alba* 24. *M. nigra* 25. *Berberis* spp. 26. *Indigofera* spp. 27. *Viburnum aconitifolium* 28. *Parrotiopsis Jacquemontiana* 29. *Platanus orientalis* 30. *Cotoneaster racemifolia* 31. *Crataegus oxyacantha* 32. *Cydonia oblonga* 33. *Eriobotrya Japonica* 34. *Malus baccatta* 35. *M. pumila* 36. *M. sylvestris* 37. *Prunus amygdalus* 38. *P. armeniaca* 39. *P. cerasifera* 40. *P. cerasus* 41. *P. cornuta* 42. *P. domestica* 43. *P. insititia* 44. *P. Jacquemontii* 45. *P. persica* 46. *P. prostrata* 47. *P. tomentosa* 48. *Pyrus communis* 49. *P. lindleyi* 50. *P. pashia* 51. *Sorbus cashmiriana* 52. *S. lanata* 53. *Amorpha fruticosa* 54. *Cercis siliquastrum* 55. *Desmondium elegans* 56. *Robinia pseudoacacia* 57. *Sophora japonica* 58. *Ailanthus altissima* 59. *Melia azedareh* 60. *Rhus succedanea* 61.*Euonymus fimbriatus* 62. *E. hamiltonianus* 63. *E. Japonicus* 64. *Acer caesium*

65. *A. gundo* 66. *Aeseulus indica* 67. *Rhamnus prostrata* 68. *R. purpurea* 69. *Zizyphus Jujuba* 70. *Vitis vinifera* 71. *Myriaria* sp. 72. *Elacagnus* sp. 73. *Hippophae* sp. 74. *Punica granata* 75. *Fraxinus excelsior* 76. *Jasminum* sp. 77. *Lingustrium* sp. 78. *Syringa* sp. 79. *Buddleja* sp. 80. *Rhododendron* sp. 81. *Lonicera* sp. 82. *Catalpa* bignonoides.

This check list includes exotic tree species. Assistance rendered by Dr. A.R. Naqshi in compiling the list is gratefully acknowledged.

INDEX

Abies 73, 76, 82, 186
Acer 75, 77, 84, 164
Achlleoin 129
Adonis aestivalis 48
Aegilopus tauschii 70
Aeschynomene indica 52
Aesculus 77, 84, 85
Aggtelek cave 132, 134, 140
Agriculture 210
    history 207
    origin 207
Ahar 112
Aldwick Barely 129
Ali Kosh 117, 126, 129
Almus 81, 94
Amaranthaceae 67
Amaranthus blitum 67, 74
A. caudatus 67, 74
A. hybridus 67, 74
A. viridis 67, 74
Ammi majus 55, 61
Amouq 129
Anantnag 3, 10
Anemone biflora 48, 56
Archaeological Survey
    of India 3, 9, 11
Archangelica officinalis 55
Arctium lappa 59
Argissa Maghula 126
A. tournefortiana 59
Artemisia scorpia 59
A. tournefortiana 59, 67
Artificial carbonization 101
Asparagus filicinus 70
A. officinalis 70
Asteracea 55, 59
Avena abyssinica 33
A. barbata 33
A. brevis 33

A. byzantina 33
A. fatua 32, 33, 35, 40, 127, 129
A. magna 128
A. murphyi 33
A. nuda 33
A. nudibrevis 33
A. sativa 30, 32, 35, 127
A. sterilis 33, 129
A. strigosa 33, 129
A. weistii 33

Banawali 126
Ban Na Di 112
Barley 28, 122, 124, 126, 130, 132
Beidha 126, 129
Ben Chiang 112
Betula 77, 84, 159
Bidens biternata 59
B. cernua 59
Biotic factor 203
Bunium persicum 55
Boraginaceae 63
Brassica compestris 6
Bromus japonicus 70
Bronze Age 129
Buplerium candollei 55
B. falcatum 55
B. tenue 55
Burzahom 6–9, 159
Buxus 161

Cajanus cajan 40
Can Hasan 126
Cannabis sativa 70, 86
Capsella bursa-pastoris 49
Carduus edelbergii 59
C. nutans 60
Carpesium abrotanoides 60

Index 275

Carpinus 84
Carum bulbocastanum 55
C. Carvi 56
Castanea 86
Catal Huyuk 126
Catalpa 86
Caucaulis latifolia 56
Causinia microcarpa 60
Cedrus 74, 82, 188
Celtis 86, 203
Cerastium glomeratum 49
Ceratocephalus falcatus 48
Cereals 14, 105
Chaerophyllum villosum 56
Chalcolithic 11
Chenopodiaceae 67
Chenopodium album 67
C. blitum 67
C. foliosum 67
C. glaucum 67
Chi square analysis 226
Chih-Shan-Yen 113
Chinar 151
Chirand 112, 138
Chopane-Mando 108
Cicer arietinum 40
Circium wallichii 60
Clematis gouriana 48
Climate 199
C. vulgare 65
Clinopodium umbrosum 65
C. vulgare 65
Co-efficient of similarity 235
Conium macolatum 56
Convolvulus arvensis 63
Coronopus didymus 49
Crataegus 198
Crepis sancta 60
Cupressus 73, 82
Cuscutta chinesis 63
Cyamopsis tetragonolobus 40
Cynoglossum glochidiatum 63
Cyperaceae 70
Cyperus iria 70
C. rotundus 70
C. serotinus 70

Dal lake 6
Datura stramonium 63
Daucus carota 56
Descurainea sophia 49
Dianthus Jacquemontii 49

Diffusion 193
Digitaria sunguinolatus 70
Dolichos biflorus 40
D. lablab 40
Drupes 42

Echinochloa crusgalli 71
Eleusine coracana 39
Elschotzia cristata 65
E. densa 66
Endocarps 42, 142
Epilobium parviflorum 55
Erigeron alpinus 60
E. bonariensis 60
E. canadensis 60
E. multicaulis 60
Eryngium billardieri 56
Euclidium syriacum 49
Euphorbiaceae 69
Euphorbia helioscopia 69
E. kanoarica 69
E. prostrata 69
E. tibetica 69

Fagopyrum esculentum 67
Ficus 163
Foeniculum vulgare 56
Forestry 212
Fraxinus excelsior 6, 86, 175
Funaria indica 48

Galium aparine 57, 147
G. asperuloides 57, 147
G. boreale 59
G. serphylloides 59
G. tenuissimum 59
G. tricorne 59, 147
G. tricolora 59
Geraniaceae 51
Geranium nepalensis 51
G. sibericum 52
Geum urbanum 53
Glume Wheats 15
Gofkral 121, 126, 140

Hacilar 126
Harappa 126, 127
Heng Chuin 119
Heracleum candicans 56
Hibiscus syriacus 51
H. trionum 51

Hieracium umbulatum 60
Hindu Rule Phase 13, 127, 142
Ho-Mu-Tu 113
Hordeum 29, 98, 122
Hordeum distichum 29, 30, 126
H. hexastichum 32
H. spontaneum 29, 124
H. vulgare 35, 98, 126,
Horticultural Fruits 37, 140
Hydrocotyle javanica 56
Hypericum perforatum 51

Inamgaon 135
Index of similarity 263
Indo-Greek 12, 127
Intensity of occupation 230
Ipomoea eriocarpa 63
I. hispida 63
I. palmata 63
Iridaceae 69
Iris ensata 69
IWGP 2

Jemdt Nasr 132
Juglans regia 38, 45, 47, 50, 86, 141, 203
J. nigra 86
Juniperus 82

Kalibangan 126
Karewa 6
Kashmir archaeological sites 4
Koldihiva 112
Kushan 12, 127

Lactuca dissecta 62
Lamiaceae 65
Lamium amplexicaule 66
Lathyrus sativus 40
Lengyel 129
Lens culinaris 40, 137
Lepidium apetalum 49
L. capitatum 49
Lespedeza cuneata 52
L. Juncea 52
L. tomentosa 52
Liliaceae 70
Linaria palmatica 63
Lithospermum arvense 63, 204
Lolium temulentum 71
Lothal 112

Luo-Jia-Jiao 113
Lycnis coroncria 51
Lycopus europalus 66

Mahadeva Hills 6
Maiden Castle 129
Malus sylvestris 6
Malva neglecta 51
Marrubium vulgare 66
Medicago lupilina 52
M. minima 52
M. polymorpha 52
M. sativa 52
Megalithic 8
Mehrgarh 122, 126, 129, 146
Melilotus albus 53, 153
Medicago indicus 53
Millets 36, 139
Mohenjodaro 121, 126
Morus 88, 153
Myriactis nepalensis 62

Naked Wheats 14
N.B.P. 11, 114, 127, 137
Nasturtium officinale 49
Neolithic 2, 8, 115, 132
Neolithic revolution 1
Nepeta cataria 66
Nirmud 132

Oats 129
Ocampo 134
Ocimum sanctum 66
Oenothera glazioviana 55
O. sovitziana 55
Onagraceae 55
Onopodium acanthuim 62
Organum vulgare 66
Oriya Timbo 134
Oryza breviligulata 108
O. glaberrim 100
O. perenis 100
O. rufipogon 108
O. sativa 30, 36, 38
Oxalis corniculata 52

Panicum 36, 132
Papaver macrostomum 48
Papilionaceae 52
Parrotiopsis 88, 168
Paspalum sacrobiculatum 34

Peganum harmala 53
Phaseolus aconitifolius 40, 135
P. aureus 40, 134
P. mungo 42, 134
Phleum paniculatum 71
Picea 73, 82, 186
Pimpinella bella 56
Pinus 7, 82, 186
Pisiim sativum 42, 138
Plantago lanceolata 66
P. major 66
Platanus 79, 88, 151, 261
Pleistocene 264
Pleistocene overkill 1
Pliocene 261
Poa bulbosa 71
P. pretensis 71
Polygonum Convoluilus 67
P. hydropiper 69
P. fugax 71
P. monospliensis 71
P. nepalense 69
Pupulus 88, 94, 203
Portulaca oleracea 51
Potentilla supina 53
Pre-N.B.P. 11
Prunus 90, 94, 182
P. amygdalus 38, 142
P. armeniaca 38, 48, 90, 203
P. cerasus 38, 48, 144
P. domestica 38, 52, 90, 142
P. persica 38, 52, 90, 143
Pulses 37, 134

Quercus 156
Quercus robur 92
Q. semicarpifolia 92, 159

Rangpur 112
Ranunculaceae 48
Ranunculus arvensis 48
R. laetus 48
R. scleratus 48
Robinia pseudoacacia 7, 92
Rorripa islandica 49
Rosaceae 53
Rosa macrophylla 53
R. webbiana 53
rubiaceae 57
Rumex nepalensis 69
Rutaceae 53

Salix 6, 92, 203
Salvia moorcroftiana 66
Scandix pecten-veneris 56
Scrophulariaceae 63
Scrophularia intermedia 65
Selinium teniufolium 56
S. wallichiana 57, 92
Semthan 10, 136
Senecio chrysanthemoides 62
S. vulgaris 62
Seseli sibiricum 57
Setaria 132
S. glauca 39, 71
S. italica 39, 132
S. viridis 39, 71
Siegesbeckia orientalis 62
Silene conoidea 51
S. dichotoma 51
Silk Route 200
Sium latijugam 57
Solanum miniatum 63
Soncluis aryensis 62
S. asper 62
S. olaraceus 62
Sorghum bicolor 39
S. halepense 71
Species diversity 229
Species evenness 229
Species richness 229
Spergularia rubra 51
Statistics 216, 263
Stone fruits 37

Tape-Guran 126
Tape-Sabz 126
Taraxacum officinale 62
Taxus 73
Tehuacan 134
Tell as Sawwan 126
Tel-Mureybit 126
Thessaly 129
Thuringa 132
Torilis japonica 57
T. leptothylla 57
T. nodosa 57
Tragopogon kashmirianus 62
Trifolium pretense 53
T. repense 53
Triticum 14
Triticum algilopoides 17, 20
T. aestivum 15, 26, 117, 121
T. araticum 15

T. boeoticum 15
T. compactum 15, 17, 110
T. dicoccoides 15, 17
T. dicoccum 15, 17
T. durum 15
T. macha 15
T. persicum 15, 17
T. polonicum 15
T. spelta 15, 17, 118
T. sphaerococcum 15, 17, 118
T. timopheevi 15
T. turgidum 15, 24
T. vavilovii 15, 110
Turgenia latifolia 57

Ulmus 6, 94, 177
Ulu-Leang 112

Valerianellia muricate 62
Vegetation 201, 263, 245
Verbascum thapsus 65
Verbena officinalis 65
Veronica anagallis-aquatica 65
V. arvensis 65
V. didyma 65

V. persica 65
Veth 10
Viburnum 94, 182
Vicia faba 42
V. sativa 53
Vicatia conifolia 57
V. wolfiana 57
Vigna sinensis 42
Vitsta 10

Wadi Kubanya 124
Weeds 49, 147
Wheat 14
Woods 73, 151
World Archaeological Congress 2

Xom Trai Cave 113

Yale Cambridge University
  Expedition Team 6
Yang Shao 113, 132

Zizyphus jujuba 38, 45
Zurich lake 134

Printed in India